Basic Concepts in
INFORMATION THEORY AND STATISTICS

Basic Concepts in INFORMATION THEORY and STATISTICS
Axiomatic Foundations and Applications

A.M. MATHAI
McGill University

P.N. RATHIE
Indian Institute of Technology, Bombay

A Halsted Press Book

JOHN WILEY & SONS
New York · London · Sydney · Toronto

Copyright © 1975, Wiley Eastern Limited
New Delhi

Published in the Western Hemisphere
by Halsted Press, a Division of
John Wiley & Sons, Inc., New York

ISBN 0-470-57622-7

Library of Congress Catalog Card Number: 74-16781

Printed in India at Thomson Press (India) Limited, Faridabad

PREFACE

In this book, several measures defining the basic concepts in Information Theory and Statistics are put on axiomatic foundations. This book is a collection of the work done by the authors and by many other workers in the field for the past several years. All the chapters contain materials mainly from research articles which are either in the press or recently published. This book is the first attempt to put fundamental concepts in Statistics and Information Theory on axiomatic foundations.

Some of the materials in this book was given as a part of a graduate course at University of Waterloo in 1969-70 and as a Ph.D course at the Indian Institute of Technology, Bombay, by the second author. Part of the book was given in a graduate course at McGill University in 1971-72 by the first author. This book is written for the graduate students and research workers in the fields of Information Theory and Foundations of Statistics and can be used as a textbook for a one semester course in these fields. This can also be used as a theoretical supplement to a course in Information Theory for engineers, economists, probabilists, statisticians and social scientists.

The book is divided into five chapters. In chapter 1 a measure of uncertainty, commonly called Shannon's entropy, and its generalizations and applications are discussed. A number of properties which are the postulates in various characterization theorems, are also listed. The characterization theorems given are for Shannon's entropy and entropies of order α. Shannon's entropy is used extensively in Information Theory, Communication Theory and in various other disciplines some of which are listed in the section on applications.

Chapter 2 deals with the concept of directed divergence, its generalizations and applications. The directed divergence and its generalizations are discussed in detail for the discrete distributions. Some other generalizations and the generalized directed divergence for three discrete distributions are also pointed out. A measure of pseudo directed divergence between two sequences $(p_1,\ldots,p_n)$ and $(q_1,\ldots,q_n)$ where $\Sigma p_i = 1$ and $\Sigma q_i \leq 1$ is also discussed. Characterization theorems are given in each case.

In chapter 3 the concept of inaccuracy, which is useful in

Statistical Inference, is discussed in detail along with its properties, generalizations, characterizations and applications.

Chapter 4 deals with the basic statistical concepts such as dispersion, covariance, affinity between distributions, discrepancy between distributions and so on, their axiomatic characterizations and some applications.

In chapter 5 some measures analogous to the ones discussed in chapters 1,2,3 and 4 are discussed. The expressions for these measures are of the same form as discussed in earlier chapters but instead of probability distributions, we have sequences of non-negative numbers where the sum of each sequence is not unity. Thus this chapter includes measures not involving probabilities. Various inequalities involving the different measures are included in this chapter as a separate section. In addition to this we also discuss some other measures useful in statistical pattern recognition.

Each chapter contains some open problems, applications and exercises. Most of the exercises are taken from recent research papers. Lots of research problems are mentioned in each chapter which will be of use to research workers in Information Theory, Foundations of Statistics, Functional Equations and related fields. Elementary results which are available in standard textbooks are not included in this book. Also we have not included measures dealing with conditional probabilities.

A few applications of the various measures in Statistics and Information Theory are also pointed out at appropriate places. Information and statistical measures involving continuous probability distributions are also discussed at some places. Many authors have asked some searching questions about "information" and have come to the conclusion that "information" is essentially undefinable. So the best solution is to search for desirable properties that an experimenter is looking for. Then put forward these properties as axioms and define the measure as the measure uniquely determined by these sets of postulates. Hence the emphasis in this book is on the mathematical foundations of the various concepts.

This book will be of use to workers in various fields because a number of properties and various characterizations of the different measures are given so that from the point of view of applications the experimenter can examine the experimental situations

Preface

corresponding to the various postulates and choose the appropriate measures. This book is an updated version of the authors' earlier monograph, "Axiomatic Foundations of Some Basic Concepts in Information Theory and Statistics" which was circulated from the Department of Mathematics, McGill University, in 1971.

Only due to the encouragement of Professor C.R.Rao the authors decided to make an attempt to put the recent research work on the foundations of Information Theory and Statistics into a book form. The authors would like to take this opportunity to express their sincere thanks to him. The authors would also like to thank the various researchers who have sent in the reprints and preprints of their research papers and especially to Professor Godfried Toussaint of the School of Computer Science at McGill University for compiling the materials on inequalities and their applications which are included in chapter 5 and to Professor Giorgio Pederzoli of Concordia University for pointing out some applications of entropy in marketing problems and for his comments on chapters 1-3.

The collaboration on this project was made possible through the National Research Council of Canada grant No.A3057. The authors acknowledge with thanks the financial support rendered by the National Research Council of Canada.

Montreal
July 1974

A.M. Mathai
P.N. Rathie

CONTENTS

Chapter 1 THE CONCEPT OF UNCERTAINTY

   1.0 Introduction   1
   1.1 The Concept of Entropy and Its Characterizations   2
   1.2 Entropies of order α 12
   1.3 Generalizations and Other Measures   17
   1.4 Continuous Analogues   23
   1.5 Applications   24
   1.6 Open Problems   28
       Exercises   28

Chapter 2 THE CONCEPT OF DIRECTED DIVERGENCE

   2.0 Introduction   35
   2.1 Directed Divergence Between Two Distributions   35
   2.2 Directed Divergence of Order α 46
   2.3 Some Generalizations   54
   2.4 Directed Divergences Involving More than Two Distributions   55
   2.5 Pseudo-Measures of Directed Divergences   61
   2.6 Continuous Analogues   67
   2.7 Applications   67
   2.8 Open Problems   69
       Exercises   70

Chapter 3 THE CONCEPT OF INACCURACY

   3.0 Introduction   75
   3.1 Inaccuracy and Its Axiomatic Characterization   75
   3.2 Inaccuracies of Order α 80
   3.3 Pseudo-Measures   83
   3.4 Further Generalizations   84
   3.5 Continuous Analogues   84
   3.6 Applications   84
   3.7 Open Problems   85
       Exercises   85

Chapter 4 SOME BASIC STATISTICAL CONCEPTS AND THEIR CHARACTERIZATIONS

4.0	Introduction	87
4.1	Covariance and Its Axiomatic Foundation	88
4.2	The Concept of Affinity and Its Characterizations	90
4.3	A Measure of Discrepancy	96
4.4	The Concept of Dispersion	97
4.5	Other Measures	101
4.6	Applications	102
4.7	Open Problems	102
	Exercises	102

Chapter 5 SOME OTHER MEASURES AND INEQUALITIES

5.0	Introduction	105
5.1	Measures Not Involving Probabilities	105
5.2	Some Generalizations	108
5.3	Measures Useful in Pattern Recognition	109
5.4	Inequalities Among Information Measures	110
5.5	Applications	114
5.6	Open Problems	115
	Exercises	115
	REFERENCES	119
	LIST OF SYMBOLS	132
	AUTHOR INDEX	134
	SUBJECT INDEX	136

Basic Concepts in
INFORMATION THEORY AND STATISTICS

CHAPTER 1

THE CONCEPT OF UNCERTAINTY

1.0 INTRODUCTION

This chapter deals with the concept of entropy. Entropy is a measure of uncertainty or a measure of information and it originated in the work of Shannon (1948). Since 1948 a number of research papers have been published which simplify and extend Shannon's original work. A number of properties and axiomatic characterizations published in the various papers to date are given either in the text or in the excercises at the end of the chapter.

Throughout this book, logarithms are taken to the base 2 and Σ stands for the summation $\Sigma_{i=1}^{n}(.)$ unless otherwise specified.

Shannon's entropy is generalized to the entropies of order α and this will be discussed in detail later on. Further generalizations of entropy are also discussed in this chapter. Applications of entropies to Coding Theory, Questionnaire Theory etc. are also pointed out. Continuous analogues are mentioned along with some open problems.

Most of the work presented in the text are taken from the research articles of Shannon (1948), Tverberg (1958), Chaundy and McLeod (1960), Campbell (1965), Jelinek (1968), Daróczy (1970), Rathie (1970,1971b,1971c,1973), Aczél, Forte and Ng (1973) ,Forte and Ng (1973) and Kannappan and Rathie (1973). The excercises are taken from the research papers of Khinchin (1953), Faddeev (1956), Renyi (1959,1961), Aczél and Daróczy (1963,1963a), Lee (1964), Kendall (1964), Campbell (1966), Pintacuda (1966a), Borges (1967), Havrda and Charvat (1967), Daróczy (1964,1967,1967a,1969a,1971), Forte and Daróczy (1968), Daróczy and Katai (1970),Rathie and

Kannappan (1971), Rathie (1972) and Forte (1973).

1.1 THE CONCEPT OF ENTROPY AND ITS CHARACTERIZATIONS

Let S_n and S_n^* denote the set of all finite discrete probability distributions $\{P \equiv (p_1,\ldots,p_n) | \ p_i \geq 0, \ i=1,\ldots,n, \ \Sigma p_i = 1\}$ and $\{P^* \equiv (p_1,\ldots,p_n) | \ p_i > 0, \ i=1,\ldots,n, \ \Sigma p_i = 1\}$ respectively. In other words, P may be regarded as an experiment having n possible outcomes with probabilities $p_1,\ldots,p_n$. Then the entropy of the distribution P, as defined by Shannon (1948), is given by

(a) DEFINITION 1.1.1

(1.1.1) $\quad H_n(p_1,\ldots,p_n) = - \Sigma \, p_i \, \log p_i$

for $P \in S_n$ and for all $n=1,2,\ldots$ In *(1.1.1)* and what follows we employ the usual convention that whenever a $p_i = 0$, $0 \log 0 = 0$.

Remark: When logarithms to base 2 or 10 or e are used, the unit of measurement is called a "bit" or "Hartley" or "nit" respectively.

A special case of *(1.1.1)* for $n=2$ which is called Shannon's entropy function and is of fundamental importance is given below.

(1.1.2) $\quad H_2(p, 1-p) = - p \log p - (1-p) \log(1-p), \quad p \in [0,1]$

Writing

(1.1.3) $\quad f(p) = H_2(p, 1-p), \ p \in [0,1]$,

where H_2 is given by *(1.1.2)*, Shannon's entropy of the distribution P is given by the expression,

(1.1.4) $\quad H_n(P) = \Sigma_{i=2}^{n} r_i \, f(p_i/r_i)$

where $r_i = p_1 + \ldots + p_i$ for all $i = 1,\ldots,n$.

(b) PROPERTIES OF ENTROPY

We will state, without proof, several interesting properties of Shannon's entropy *(1.1.1)* and the entropy function *(1.1.2)*. Possible interpretations of these properties are also given. Similar interpretations can be given to other measures defined in this book.

(i) Non-negativity:

(1.1.5) $\quad H_n(P) \geq 0$.

The Concept of Entropy and Its Characterizations

with equality if and only if one $p_i = 1$ and the rest are zeros.

(ii)   Expansibility or zero-indifferent

(1.1.6) $H_{n+1}(P,0) = H_n(P)$

Expansibility means that the entropy does not change if we add an outcome or outcomes with zero probability.

(iii)   Symmetry

(1.1.7) $H_n(p_1,\ldots,p_n) = H_n(p_{a_1},\ldots,p_{a_n})$

where $\{a_1,\ldots,a_n\}$ is an arbitrary permutation of $\{1,\ldots,n\}$ for all $P \in S_n$ and $n=1,2,\ldots$. This says that the entropy does not depend on the order in which the possible outcomes are labeled.

(iv)   Recursivity or Branching Principle: For $P \in S_n$, $n=3,4,\ldots$, $p_1+p_2 > 0$

(1.1.8) $H_n(p_1,\ldots,p_n) = H_{n-1}(p_1+p_2, p_3,\ldots,p_n)$
$$+ (p_1+p_2) H_2(p_1/(p_1+p_2), p_2/(p_1+p_2))$$

This property says that if a choice is broken down into two successive choices, the original H should be the weighted sum of the individual values of H.

(v)   Additivity: For $P \in S_n$ and $Q \in S_m$,

(1.1.9) $H_{mn}(p_1q_1,\ldots,p_1q_m,\ldots,p_nq_1,\ldots,p_nq_m) = H_n(p_1,\ldots,p_n) + H_m(q_1,\ldots,q_m)$

This property says that, in case of independent experiments, the entropy of the combination of two experiments is equal to the sum of the entropies of the individual experiments.

(vi)   Strong Additivity: For $p_{ij} \geq 0$, $p_j = \sum_{i=1}^{n} p_{ij} > 0$, $j=1,\ldots,m$, $\sum_{i=1}^{n} \sum_{j=1}^{m} p_{ij} = 1$, we have

(1.1.10) $H_{mn}(p_{11},\ldots,p_{1m},\ldots,p_{n1},\ldots,p_{nm}) = H_m(p_1,\ldots,p_m)$
$$+ \sum_{j=1}^{m} p_j H_n(p_{1j}/p_j,\ldots,p_{nj}/p_j)$$

(vii)   Continuity: $H_n(P)$ is a continuous function of its n variables. To some extent, this implies that when the probabilities are changed slightly there is a corresponding slight change in the entropy.

(viii)   Monotonicity: $H_n(1/n,\ldots,1/n)$ is a monotonic increasing function of n.

(ix) Inequality

(1.1.11) $\quad H_n(P) \leq H_n(1/n,\ldots,1/n)$

with equality if and only if $p_i = 1/n$ for all $i=1,\ldots,n$. This property states that the maximum of the entropy is attained when all the probabilities are equal to $1/n$.

(x) Sub-additivity: For $\sum_{i=1}^{m} \sum_{j=1}^{n} p_{ij} = 1$, $p_{ij} \geq 0$ for all i,j

(1.1.12) $\quad H_{mn}(p_{11},\ldots,p_{1n},\ldots,p_{m1},\ldots,p_{mn}) = H_m(\sum_{j=1}^{n} p_{1j},\ldots,\sum_{j=1}^{n} p_{mj})$
$$+ H_n(\sum_{i=1}^{m} p_{i1},\ldots,\sum_{i=1}^{m} p_{in})$$

This inequality says that the entropy of a combination of two experiments can not exceed the sum of the entropies of the single experiments.

(xi) Functional Equation: f defined in *(1.1.3)* satisfies the functional equation,

(1.1.13) $\quad f(x) + (1-x)f(y/(1-x)) = f(y) + (1-y)f(x/(1-y))$

where $x,y \in [0,1[$ with $x+y \in [0,1]$ and the boundary conditions $f(0) = f(1)$ and $f(\tfrac{1}{2}) = 1$.

More properties of entropy may be found in the exercises at the end of this chapter.

(c) CHARACTERIZATION THEOREMS

Shannon's entropy *(1.1.1)* has been characterized by different sets of postulates by various workers in the field. We will be discussing below a few characterization theorems in detail and the rest will be put in the exercises at the end of this chapter.

The first theorem which is of historical importance is due to Shannon (1948). The postulates given by Shannon (1948) are incomplete and in order to get the characterization theorem, we must also assume symmetry as one of the postulates. Here we are presenting a modified version of Shannon's results by assuming the following conditions as reasonable properties of entropy.

A1. Continuity: Property *(iii)* of §1.1.
A2. Monotonicity: Property *(viii)* of §1.1, namely,

(1.1.14) $\quad H_n(1/n,\ldots,1/n) \leq H_{n+1}(1/(n+1),\ldots,1/(n+1))$

A3. *Recursivity:* If a choice be broken down into two successive choices, the original H should be the weighted sum of the individual values of H. We may translate this to mathematical language as follows.

(1.1.15) $\quad H_n(p_1,\ldots,p_{m-1},p_m q_1, p_m q_2,\ldots,p_m q_{n-m+1}) = H_m(p_1,\ldots,p_m)$
$$+ p_m H_{n-m+1}(q_1,\ldots,q_{n-m+1})$$

where $\Sigma_{i=1}^{m} p_i = 1$, $\Sigma_{i=1}^{n-m+1} q_i = 1$.

A4. *Normalization*

(1.1.16) $\quad H_2(\tfrac{1}{2},\tfrac{1}{2}) = 1$

A5. *Symmetry:* (1.1.7) of §1.1.

THEOREM 2.1.1: The only H_n satisfying the five postulates A_1 to A_5 is Shannon's entropy (1.1.1).

Proof: Let

(1.1.17) $\quad H_n(1/n,\ldots,1/n) = F(n)$

From postulate A_3 and A_5 we can breakdown a choice from s^m equally likely possibilities into a series of m choices each from s equally likely possibilities and get

(1.1.18) $\quad F(s^m) = m F(s)$

Similarly
$$F(t^n) = t F(t)$$

Choose n arbitrarily large and find an m to satisfy

(1.1.19) $\quad s^m \leqslant t^n \leqslant s^{m+1}$

Thus, taking logarithms in (1.1.19) and dividing by $n \log s$, one gets

$$m/n \leqslant (\log t)/(\log s) \leqslant m/n + 1/n$$

or

(1.1.20) $\quad |m/n - (\log t)/(\log s)| < \varepsilon$

where ε is an arbitrary small constant.

Also, applying postulate A_2 to (1.1.19) and utilizing (1.1.18) we have

(1.1.21) $\quad m F(s) \leqslant n F(t) \leqslant (m+1) F(s)$

Now, as before, (1.1.21) on dividing by $nF(s)$ gives,

(1.1.22) $\quad |m/n - F(t)/F(s)| < \varepsilon$

Hence, from (1.1.10) and (1.1.12), we have

$$|F(t)/F(s) - (\log t)/(\log s)| \leq 2\varepsilon$$

giving

(1.1.23) $\quad F(t) = k \log t$

Clearly, $k \geq 0$ due to postulate A2.

Now suppose we have a choice from n possibilities with probabilities $p_i = n_i/\Sigma n_i$ where the n_i are integers. We can decompose a choice from Σn_i possibilities into a choice from n possibilities with probabilities $p_1, \ldots, p_n$. Using postulate A3, we equate the total choice from Σn_i as computed by two methods to get,

$$k \log \Sigma n_i = H_n(P) + k \Sigma p_i \log n_i$$

Hence,

(1.1.24) $\quad H_n(P) = -k \Sigma p_i \log(n_i/\Sigma n_i) = -k \Sigma p_i \log p_i$

The same result holds if p_i are incommensurable because of the continuity postulate A1. Applying postulate A4 to (1.1.24) we easily see that $k=1$. This proves theorem 1.1.1.

The following theorem is due to Tverberg (1958) and it is an improvement over the theorems due to Shannon (1948), Khinchin (1953) and Faddeev (1956). Let a function $H_n(P)$ for $P \varepsilon S_n$ satisfy the following postulates.

B1. *Symmetry:* Property (iii) of §1.1
B2. *Recursivity:* $H_{n+1}[p_1, \ldots, p_{n-1}, p_n q, p_n(1-q)] = H_n(p_1, \ldots, p_n) + p_n H_2(q, 1-q)$, where $q \varepsilon [0,1]$.
B3. *Regularity:* $H_2(p, 1-p)$ is Lebesgue integrable in $0 \leq p \leq 1$
B4. *Normalization:* $H_2(\frac{1}{2}, \frac{1}{2}) = 1$.

Note. We have modified the second postulate of Tverberg (1958) in B2 in order to accommodate $p_i = 0$ for any i. If we take Tverberg's postulates as such then we will have to include expansibility postulate (ii) in order that any of the p_i's may take zero value.

THEOREM 1.1.2: A function $H_n(P)$, $P \varepsilon S_n$ satisfying the postulates B1 to B4 is uniquely determined as Shannon's entropy given in (1.1.1).

Proof: Taking $p_n q = u$ and $p_n(1-q) = v$ in postulate B2, we have

(1.1.25) $\quad H_{n+1}(p_1, \ldots, p_{n-1}, u, v) = H_n(p_1, \ldots, p_n) + p_n H_2(u/p_n, v/p_n)$

The form given in (1.1.25) will be used whenever $p_i \neq 0$ for all $i = 1, \ldots, n$.

Clearly, from postulates B1, B2 and the equation (1.1.25) we have

(1.1.26) $\quad H_3(x,u,v) = H_2(x,u+v) + (u+v)H_2(u/(u+v), v/(u+v)) = H_3(u,x,v)$

$\quad\quad\quad\quad = H_2(u,x+v) + (x+v)H_2(x/(x+v),v/(x+v))$

for $x, u \in [0,1[, x+u \in [0,1]$.

Taking

(1.1.27) $\quad f(x) = H_2(x, 1-x)$, for $x \in [0,1]$

the equation (1.1.26) takes the following form:

(1.1.28) $\quad f(x) + (1-x)f(u/(1-x)) = f(u) + (1-u)f(x/(1-u))$

for $x, u \in [0,1[, x+u \in [0,1]$.

Integrating (1.1.28) with respect to u between the limits 0 to $1-x$, which is valid due to postulate B3, we get

(1.1.29) $\quad (1-x)f(x) + (1-x)^2 \int_0^1 f(t)dt = \int_0^{1-x} f(t)dt + x^2 \int_x^1 t^{-3} f(t)dt$

The continuity in x for $x \in]0,1[$ of all terms of (1.1.29) except the first one is assured due to postulate B3. Hence $f(x)$ is continuous. Therefore, by an analogous argument $f(x)$ is differentiable for $x \in]0,1[$. Differentiating (1.1.29), we have

(1.1.30) $\quad (1-x)f'(x) - f(x) - 2(1-x) \int_0^1 f(t)dt$

$\quad\quad\quad = -f(1-x) + 2x \int_x^1 t^{-3} f(t)dt - x^{-1} f(x)$

From (1.1.27) and postulate B1, we get

(1.1.31) $\quad f(x) = f(1-x)$ for $x \in [0,1]$

Now cancelling $f(x)$ and $f(1-x)$ in (1.1.30) we observe, as before, that $f''(x)$ exists. Thus differentiating (1.1.30) and then eliminating $\int_x^1 t^{-3} f(t)dt$, we get

(1.1.32) $\quad f''(x) = -2x^{-1}(1-x)^{-1} \int_0^1 f(t)dt$

Hence (1.1.32) gives

(1.1.33) $\quad f(x) = ax + b - 2[x \log_e x + (1-x)\log_e(1-x)] \int_0^1 f(t)dt$

Clearly (1.1.33) and (1.1.31) show that $a=0$. Also integrating (1.1.33) with $a=0$ from 0 to 1 gives $b=0$. Thus

(1.1.34) $\quad f(x) = -2 \int_0^1 f(t)dt \, [x \log_2 x + (1-x)\log_2(1-x)] \log_e 2$

for $x \in]0,1[$. Now postulate B4 and (1.1.34) yield

$\quad\quad 2 \int_0^1 f(t)dt \, \log_e 2 = 1$

giving

(1.1.35) $f(x) = - x \log x - (1-x)\log(1-x)$, $x \in \,]0,1[$

Taking $u = 1-x$ in (1.1.28) and utilizing (1.1.31), one gets $f(1) = 0$ which on using (1.1.31) yields

(1.1.36) $f(0) = f(1) = 0$

Hence (1.1.35) and (1.1.36) give $f(x)$, that is, $H_2(x,1-x)$ is obtained for $x \in [0,1]$. By induction and by the use of postulates B1 and B2, one gets

(1.1.37) $H_n(P) = \sum_{i=2}^{n} r_i \, f(p_i/r_i)$

where $r_i = p_1 +...+ p_i$ for all $i=1,...,n$. Hence (1.1.35) and (1.1.37) yield

$$H_n(P) = - \sum_{i=2}^{n} r_i [(p_i/r_i)\log(p_i/r_i) + (1-p_i/r_i)\log(1-p_i/r_i)]$$

$$= - \sum_{i=2}^{n} p_i \log p_i - \sum_{i=2}^{n} r_i \log(r_{i-1}/r_i) + \sum_{i=2}^{n} p_i \log r_{i-1}$$

$$= - \sum_{i=2}^{n} p_i \log p_i + \sum_{i=2}^{n} r_i \log r_i - \sum_{i=2}^{n} r_{i-1} \log r_{i-1}$$

$$= - \sum_{i=2}^{n} p_i \log p_i - r_1 \log r_1 = - \sum_{i=1}^{n} p_i \log p_i$$

This completes the proof of theorem 1.1.2.

The theorem given below is essentially due to Chaundy and McLeod (1960). Let the functions $H_n(P)$, $P \in S_n$ and $f(.)$ satisfy the following postulates.

C1. *Ordinary Sum:* $H_n(P) = \sum_{i=1}^{n} f(p_i)$
C2. *Continuity:* $f(.)$ is continuous in $[0,1]$
C3. *Additivity:* Property (v) of §1.1
C4. *Normalization:* $H_2(\frac{1}{2},\frac{1}{2}) = 1$

Then we have the following theorem.

THEOREM 1.1.3: The function $H_n(P)$ having the representation as given by the postulate C1 and satisfying the postulates C2 to C4 is Shannon's entropy (1.1.1).

Proof: From postulates C1 and C3, we have

(1.1.38) $\sum_{i=1}^{n} \sum_{j=1}^{m} f(p_i q_j) = \sum_{i=1}^{n} f(p_i) + \sum_{j=1}^{m} f(q_j)$

The Concept of Entropy and Its Characterizations

for $P \varepsilon S_n$ and $Q \varepsilon S_m$. Now we will show that

(1.1.39) $\qquad f(x) = Ax \log x, \quad x \varepsilon [0,1]$

where A is a constant. Let p,q,r,s be any integers such that $1 \leq r \leq p$, $1 \leq s \leq q$. Taking $n = p-r+1$, $m = q-s+1$, $p_i = 1/p$, $i = 1,\ldots,p-r$, $p_{p-r+1} = r/p$, $q_j = 1/q$, $j = 1,\ldots,q-s$ and $q_{q-s+1} = s/q$, in (1.1.39), one gets

(1.1.40) $\qquad (p-r)(q-s)f(1/pq) + (p-r)f(s/pq) + (q-s)f(r/pq) + f(rs/pq)$

$\qquad\qquad = (p-r)f(1/p) + f(r/p) + (q-s)f(1/q) + f(s/q)$

Multiplying (1.1.40) by pq and putting

(1.1.41) $\qquad x f(1/x) = \phi(x)$

we get

(1.1.42) $\qquad (p-r)(q-s)\phi(pq) + s(p-r)\phi(pq/s) + r(q-s)\phi(pq/r) + rs\phi(pq/rs)$

$\qquad\qquad = q(p-r) \phi(p) + qr \phi(p/r) + p(q-s) \phi(q) + ps \phi(q/s)$

Taking $r=s=1$ in (1.1.42), we get

(1.1.43) $\qquad \phi(pq) = \phi(p) + \phi(q)$

Again, taking $s = 1$ in (1.1.42), we have

$\qquad q(p-r) \phi(pq) + qr \phi(pq/r) = q(p-r) \phi(p) + qr \phi(p/r) + pq \phi(q)$

which, on substitution for $\phi(pq)$ from (1.1.43), yields

$\qquad \phi(pq/r) = \phi(p/r) + \phi(q)$

Also by symmetry,

$\qquad \phi(pq/s) = \phi(p) + \phi(q/s)$

Now substituting for $\phi(pq)$, $\phi(pq/r)$ and $\phi(pq/s)$, in (1.1.41), we obtain

$\qquad \phi(pq/rs) = \phi(p/r) + \phi(q/s)$

That is,

(1.1.44) $\qquad \phi(xy) = \phi(x) + \phi(y)$

where $x,y \geq 1$ and x,y are rational. Due to the continuity postulate

$C2$, the equation $(1.1.44)$ holds good for all $x, y \geq 1$. The equation $(1.1.44)$ is the well-known Cauchy equation of which the general continuous solution is

$(1.1.45) \qquad \emptyset(x) = - A \log x$

where A is an arbitrary constant. Hence, $(1.1.45)$ and $(1.1.41)$ give $f(x) = A\, x \log x$, that is $(1.1.39)$. It is easy to verify that $(1.1.39)$ satisfies $(1.1.38)$.

Now postulate $C1$ and $(1.1.39)$ yield

$(1.1.46) \qquad H_n(p_1,\ldots,p_n) = A\ \Sigma p_i \log p_i$

Applying postulate $C4$ to $(1.1.46)$ for $n=2$ one gets $A=-1$, thus proving theorem $1.1.3$.

The next two characterization theorems are due to Aczél, Forte and Ng (1974).

THEOREM 1.1.4: The functions $K_n(p_1,\ldots,p_n) : S_n \to \mathcal{R}$ $(n=2,3,\ldots)$ satisfying the postulates

D1. *Expansibility*: $(1.1.6)$ of §1.1
D2. *Symmetry*: $(1.1.7)$ of §1.1
D3. *Additivity*: $(1.1.9)$ of §1.1
D4. *Sub-additivity*: $(1.1.12)$ of §1.1, or
D4'. *Weak Sub-additivity*:

$(1.1.47) \qquad K_{2n}(p_{11},p_{12},p_{21},p_{22},\ldots,p_{n1},p_{n2}) \leq K_n(p_{11}+p_{12},\ p_{21}+p_{22},\ldots,p_{n1}+p_{n2})$
$\qquad\qquad + K_2(p_{11}+p_{21}+\ldots+p_{n1},\ p_{12}+p_{22}+\ldots+p_{n2})$

for $(p_{11},p_{12},\ldots,p_{n1},p_{n2}) \in S_{2n}$, $n = 2,3,\ldots$ are uniquely determined by the expression

$(1.1.48) \qquad K_n(p_1,\ldots,p_n) = a\, H_n(p_1,\ldots,p_n) + b \log N$

for $P \in S_n$, $n = 2,3,\ldots$, where a and b are arbitrary non-negative constants, H_n is the Shannon's entropy $(1.1.1)$ and N is the number of zeros among $(p_1,\ldots,p_n)$.

Proof: The proof of theorem $1.1.4$ depends on the following lemmas which we are mentioning below without proofs.

LEMMA 1.1.1: If K_n satisfies the postulates $D2$, $D3$ for $n = 2$ and $(1.1.47)$, then

The Concept of Entropy and Its Characterizations

(1.1.49) $K_2(1-q,q) - K_2[(1-p)(1-q) + p(1-r), (1-p)q + pr]$

$\leq K_n[p(1-q), pq, p_3, p_4, \ldots, p_n] - K_n[p(1-r), pr, p_3, p_4, \ldots, p_n]$

$\leq K_2[p(1-q) + (1-p)(1-r), pq + (1-p)r] - K_2(1-r,r)$

for all $q \in [0,1]$, $r \in [0,1]$ and all $(p, p_3, \ldots, p_n) \in S_{n-1}$ $(n = 3, 4, \ldots)$

LEMMA 1.1.2: The postulate D2 for $n = 2$ and (1.1.49) imply that the function $f(x) \equiv K_2(1-x,x)$, $x \in [0,1]$ has the following properties.

(a_1) $f(1-q) = f(q)$, $q \in [0,1]$
(a_2) f is monotonic non-decreasing on $[0,\frac{1}{2}]$ (non-increasing on $[\frac{1}{2},1]$)
(a_3) f is continuous on $]0,1[$
(a_4) f is concave on $[0,1]$, that is,

$$f[(1-\lambda)q + \lambda r] \geq (1-\lambda)f(q) + \lambda f(r) \text{ for all } \lambda, q, r \in [0,1]$$

(a_5) The right and left derivatives D^+f and D^-f exist everywhere on $[0,1[$ and $]0,1]$ respectively. Also D^+f and D^-f are finite on $]0,1[$, $D^+f(x) \geq 0$ for all $x \in [0,\frac{1}{2}[$ and $D^-f(x) \geq 0$ for all $x \in]0,\frac{1}{2}]$.

LEMMA 1.1.3: From the postulate D2 for $n = 2$ and (1.1.49) it follows that

$$K_n[p(1-q), pq, p_3, \ldots, p_n] = p\, K_2(1-q,q) + J_{n-1}(p, p_3, \ldots, p_n)$$

for all $q \in]0,1[$, $(p, p_3, \ldots, p_n) \in S_{n-1}$, $n = 3, 4, \ldots$, where $J_{n-1}: S_n \to \mathbb{R}$.

LEMMA 1.1.4: K_n satisfies the postulates D2 with $n = 2, 3$, D3 with $n = 2$ and (1.1.47), then there exists constants $a \geq 0$ and c such that

$$K_2(1-q,q) = a\, H_2(1-q,q) + c \text{ for all } q \in]0,1[$$

where H_2 is given by (1.1.1).

LEMMA 1.1.5: If K_n satisfies the postulates D2, D3 with $n = 2$ and (1.1.47), then there exist constants $a \geq 0$ and $A(n)$ $(n = 2, 3, \ldots)$ such that

(1.1.50) $K_n(p_1, p_2, \ldots, p_n) = a\, H_n(p_1, p_2, \ldots, p_n) + A(n)$

for all $(p_1, p_2, \ldots, p_n) \in S_n^*$ and all $n = 2, 3, \ldots$, where H_n $(n = 2, 3, \ldots)$ is the Shannon's entropy (1.1.1).

LEMMA 1.1.6: Under the conditions of the theorem *1.1.4* the function A of lemma *1.1.5* is given by

(1.1.51) $A(n) = b \log_2 n$ for all $n = 2, 3, \ldots$

where b is a non-negative constant.

The results *(1.1.50)* and *(1.1.51)* give

(1.1.52) $K_n(p_1, p_2, \ldots, p_n) = a H_n(p_1, p_2, \ldots, p_n) + b \log_2 n$

for $(p_1, p_2, \ldots, p_n) \in S_n^*$, $n = 2, 3, \ldots$, and $a, b \geq 0$.

Thus we have proved the theorem *1.1.4* for S_n^*, $n = 2, 3, \ldots$ It can be easily extended for S_n for $n = 2, 3, \ldots$. As a special case of the theorem *1.1.4* we deduce the following theorem for Shannon's entropy *(1.1.1)*.

THEOREM 1.1.5: Shannon's entropy *(1.1.1)* is the only entropy which satisfies the postulates *D1, D2, D3, (1.1.47)* and the conditions $K_2(\frac{1}{2}, \frac{1}{2}) = 1$, $\lim_{p \to 0^+} K_2(1-p, p) = 0$.

Proof: The property $\lim_{p \to 0^+} K_2(1-p, p) = 0$ implies $b = 0$ in *(1.1.48)* and $K_2(\frac{1}{2}, \frac{1}{2}) = 1$ gives $a = 1$ proving the theorem *1.1.5*. For further details, see the paper of Aczél, Forte and Ng (1974).

1.2 ENTROPIES OF ORDER α

This section deals with two generalizations, involving one parameter, of Shannon's entropy.

(a) DEFINITIONS

For the discrete distribution $P \in S_n$ the two types of entropies of order α are defined as follows:

DEFINITION 1.2.1: Additive entropy of order α (Rényi 1961)

(1.2.1) $\hat{H}_{n,\alpha}(P) = (1-\alpha)^{-1} \log(\Sigma p_i^\alpha)$, $\alpha \neq 1$

DEFINITION 1.2.2: Non-additive entropy of order α (Havrda and Charvát 1967)

(1.2.2) $H_{n,\alpha}(P) = (\Sigma p_i^\alpha - 1)/(2^{1-\alpha} - 1)$, $\alpha \neq 1$

Both *(1.2.1)* and *(1.2.2)* reduce to Shannon's entropy *(1.1.1)* as $\alpha \to 1$. In *(1.2.1)* and *(1.2.2)* the usual convention $0^\alpha = 0 (\alpha \neq 0)$ will be used. $\hat{H}_{n,\alpha}$ and $H_{n,\alpha}$ are connected by the relation,

Entropies of Order α

(1.2.3) $H_{n,\alpha} = [2^{(1-\alpha)\hat{H}_{n,\alpha}} - 1]/(2^{1-\alpha} - 1)$, $\alpha \neq 1$

An interesting special case of *(1.2.2)* for $n = 2$ is given below.

(1.2.4) $H_{2,\alpha}(p, 1-p) = [p^\alpha + (1-p)^\alpha - 1]/(2^{1-\alpha} - 1)$, $\alpha \neq 1$

Let

(1.2.5) $f_\alpha(p) = H_{2,\alpha}(p, 1-p)$, $p \in [0,1]$

Then we will call $f_\alpha(p)$ as the *non-additive entropy function of order* α and it is clearly a generalization of $f(p)$ given in *(1.1.3)* in the sense that as $\alpha \to 1$, $f_\alpha(p) \to f(p)$. Alternately, we may define $f_\alpha(p)$, the entropy function of order α and $H_{n,\alpha}$, the non-additive entropy of order α as follows.

DEFINITION 1.2.3: A real valued function f_α defined in $[0,1]$ will be called the non-additive entropy function of order α(α≠1) if it satisfies the functional equation

(1.2.6) $f_\alpha(x) + (1-x)^\alpha f_\alpha(y/(1-x)) = f_\alpha(y) + (1-y)^\alpha f_\alpha(x/(1-y))$

for all $x, y \in [0,1[$ with $x+y \in [0,1]$, and the boundary conditions,

(1.2.7) $f_\alpha(0) = f_\alpha(1)$

and

(1.2.8) $f_\alpha(\frac{1}{2}) = 1$

Later on it will be shown that *(1.2.5)* is the only solution of *(1.2.6)* under the conditions *(1.2.7)* and *(1.2.8)*.

DEFINITION 1.2.4: If f_α is the non-additive entropy function of order α(α ≠ 1) as given in definition *1.2.3* above, then we define the non-additive entropy of order α(α≠1) by the expression

(1.2.9) $H_{n,\alpha}(P) = \sum_{i=2}^{n} r_i^\alpha f_\alpha(p_i/r_i)$

where $r_i = p_1 + \ldots + p_i$ for all $i = 1, \ldots, n$. It is easy to verify with the help of *(1.2.5)*, that the right hand side of *(1.2.9)* is the same as the right hand side of *(1.2.2)*.

Now we will discuss the various properties and some interesting characterization theorems regarding the entropies of order α.

(b) PROPERTIES

(i) The properties *(i),(ii),(iii),(vii)* for Shannon's entropy also hold for $H_{n,\alpha}$ and $\hat{H}_{n,\alpha}$. The additive property *(v)* holds for $\hat{H}_{n,\alpha}$. The properties *(viii)* and *(ix)* hold for $H_{n,\alpha}$ for $\alpha > 0$.

(ii) *Recursivity:* For $p_1 + p_2 > 0$ and for all $n = 3, 4, \ldots$

$$(1.2.10) \quad H_{n,\alpha}(p_1,\ldots,p_n) = H_{n-1,\alpha}(p_1+p_2, p_3, \ldots, p_n)$$
$$+ (p_1+p_2)^\alpha H_{2,\alpha}(p_1/(p_1+p_2), p_2/(p_1+p_2))$$

(iii) *Non-additivity:* For $P \in S_n$ and $Q \in S_m$,

$$(1.2.11) \quad H_{mn,\alpha}(p_1 q_1, \ldots, p_1 q_m, \ldots, p_n q_1, \ldots, p_n q_m) = H_{n,\alpha}(p_1, \ldots, p_n)$$
$$+ H_{m,\alpha}(q_1, \ldots, q_m) + (2^{1-\alpha} - 1) H_{n,\alpha}(p_1, \ldots, p_n) H_{m,\alpha}(q_1, \ldots, q_m)$$

(iv) *Strong non-additivity:* For $q_{ji} \geq 0$, $\sum_{i=1}^{n} q_{ji} = 1$ for all $j = 1, \ldots, m$, $p_i \geq 0$, $\sum_{i=1}^{m} p_i = 1$, $\alpha > 0$,

$$(1.2.12) \quad H_{mn,\alpha}(p_1 q_{11}, \ldots, p_1 q_{1n}, \ldots, p_n q_{m1}, \ldots, p_n q_{mn})$$
$$= H_{m,\alpha}(p_1, \ldots, p_n) + \sum_{j=1}^{m} p_j^\alpha H_{n,\alpha}(q_{j1}, \ldots, q_{jn})$$

(c) CHARACTERIZATION THEOREMS

In this section a few of the characterization theorems will be discussed in detail and the rest will be put as exercises at the end of this chapter. First we will prove the following theorem (Daróczy 1970) for the non-additive entropy function of order α.

THEOREM 1.2.1: If f_α is the non-additive entropy function of order $\alpha (\alpha \neq 1)$ as given by the definition *1.2.3* then f_α is uniquely determined and it is given by *(1.2.5)*.

Proof: Taking $x=0$ in *(1.2.6)*, we find that $f(0) = 0$. Thus from *(1.2.7)* we get

$$(1.2.13) \quad f_\alpha(0) = f_\alpha(1) = 0$$

Taking $y = 1-x$ in *(1.2.6)* and utilizing *(1.2.13)*, we have

$$(1.2.14) \quad f_\alpha(x) = f_\alpha(1-x), \quad x \in [0,1]$$

Let $p, q \in]0,1[$ be two arbitrary numbers. Putting $p = 1-x$ and $q = y/(1-x)$ in *(1.2.6)* and using *(1.2.14)*, we get

Entropies of Order α

(1.2.15) $f_\alpha(p) + p^\alpha f_\alpha(q) = f_\alpha(pq) + (1-pq)^\alpha f((1-p)/(1-pq))$

Let

(1.2.16) $F_\alpha(p,q) = f_\alpha(p) + [p^\alpha + (1-p)^\alpha] f(q)$

Then we will show that

(1.2.17) $F_\alpha(p,q) = F_\alpha(q,p)$

The equations (1.2.16) and (1.2.15) give us

(1.2.18) $F_\alpha(p,q) = f_\alpha(pq) + (1-pq)^\alpha [f_\alpha((1-p)/(1-pq)) + ((1-p)/(1-pq))^\alpha f_\alpha(q)]$

for all $p,q \in]0,1[$. Now, applying (1.2.15) and (1.2.14) to (1.2.18) we get

$F_\alpha(p,q) = f_\alpha(pq) + (1-pq)^\alpha [f_\alpha((1-p)q/(1-pq)) + ((1-q)/(1-pq))^\alpha f_\alpha(p)]$

$= f_\alpha(pq) + (1-pq)^\alpha [f_\alpha((1-q)/(1-pq)) + ((1-q)/(1-pq))^\alpha f_\alpha(p)]$

$= F_\alpha(q,p)$

which is precisely (1.2.17).

Now (1.2.16) and (1.2.17) for $q=\frac{1}{2}$ with the help of (1.2.8) prove theorem 1.2.1 for $p \in]0,1[$ which in conjunction with (1.2.13) establish the result completely.

Next we will prove a characterization theorem for Shannon's entropy (1.1.1) and the non-additive entropy of order α (1.2.2). This result is due to Forte and Ng (1973).

THEOREM 1.2.2: If the function $K_n(p_1,\ldots,p_n)$ satisfies the postulates,

E1. *Symmetry*: $K_6(p_1,\ldots,p_6)$ is a symmetric function of its variables,

E2. *Expansibility*: $K_{n+1}(p_1,\ldots,p_n,0) = K_n(p_1,\ldots,p_n)$, $(p_1,\ldots,p_n) \in S_n$, $n \geq 2$,

E3. *Branching*: $K_{n+1}(p_1,p_2,\ldots,p_{n+1}) - K_n(p_1+p_2,p_3,\ldots,p_{n+1}) = \Delta_n(p_1,p_2) > 0$
for all $(p_1,p_2,\ldots,p_{n+1}) \in S_{n+1}$, $p_1,p_2 > 0$, $n \geq 2$,

E4. *Compositivity*: $K_6(pp_1,pp_2,pp_3,pp_4,(1-p)q_1,(1-p)q_2) = \Psi_{4,2}[K_4(p_1,p_2,p_3,p_4), K_2(q_1,q_2),p]$ for all $(p_1,p_2,p_3,p_4) \in S_4$, $(q_1,q_2) \in S_2$, $p \in [0,1]$, K_2 is not constant on S_2,

E5. *Continuity*: K_3 is continuous at the boundary points of S_3,

E6. *Nullity*: $K_2(0,1) = 0$,

E7. *Normalization*: $K_2(\frac{1}{2},\frac{1}{2}) = 1$,

then $K_n(p_1,\ldots,p_n) \equiv H_{n,\alpha}(P)$, defined by (1.2.2), which is equal to $H_n(P)$ for $\alpha=1$ defined by (1.1.1).

The proof of the theorem is based on the following lemmas which are stated here without proofs. These lemmas are under the postulates *E1* to *E5* given above.

LEMMA 1.2.1: $K_4(pq_1, p(1-q_1), (1-p)q_2, (1-p)(1-q_2))$ is continuous in p at 0.

LEMMA 1.2.2: $K_2(p, 1-p)$ is continuous on $[0,1]$.

LEMMA 1.2.3: $K_4(pq_1, p(1-q_1), (1-p)q_2, (1-p)(1-q_2))$ is continuous in both the variables $q_1, q_2 \in [0,1]$.

LEMMA 1.2.4: The range of $p \to K_2(p, 1-p)$ is a proper closed interval $[a,b]$.

LEMMA 1.2.5: $K_3(pq, p(1-q), 1-p)$ is continuous on $[0,1]$ for every fixed $q \in [0,1]$.

LEMMA 1.2.6: The function $\emptyset(u,p)$ defined by

$$\emptyset(u,p) = \Psi_{4,2}(u,k,p) - \tfrac{1}{2} K_2(p, 1-p)$$

where $k = K_2(0,1)$ has the following properties:

(b_1) $\emptyset(u,p)$ is continuous on $[0,1]$ for each $u \in [a,b]$,
(b_2) $\emptyset(u,p)$ is continuous on $[a,b]$ for each $p \in [0,1]$,
(b_3) $\emptyset(u,0) = k/2$ and $\emptyset(u,1) = u-k/2$, for every $u \in [a,b]$,
(b_4) $\emptyset(u,p) + \emptyset(v,1-p) = \Psi_{4,2}(u,v,p)$, for every $u \in K_4(S_4)$, $v \in [a,b]$, $p \in [0,1]$,
(b_5) $\emptyset\{\emptyset(u,(1-q)/p) + \emptyset(v,(p+q-1)/p), p\} + \emptyset(w, 1-p) = \emptyset(u, 1-q)$
$\qquad + \emptyset\{\emptyset(v,(p+q-1)/q) + \emptyset(w,(1-p)/q), q\}$

for every $u,v,w \in [a,b]$ and $p,q \in \,]0,1]$ such that $p+q \geq 1$.

LEMMA 1.2.7: If $\emptyset$ satisfies (b_1) to (b_5) of lemma *1.2.6*, then there exist functions $\mu, \lambda : [0,1] \to \mathcal{R}$ such that
(c_1) $\emptyset(t,p) = \mu(p)t + \lambda(p)$ for $t \in [a,b]$, $p \in [0,1]$ and the functions μ and λ satisfy:
(c_2) $\mu(0) = 0$, $\mu(1) = 1$
(c_3) $\mu(p)\,\mu((1-q)/p) = \mu(1-q)$ for $p \in \,]0,1]$, $q \in [1-p, 1[$,
(c_4) $\lambda(0) = k/2$, $\lambda(1) = -k/2$,
(c_5) $\mu(p)\{\lambda((1-q)/p) + \lambda((p+q-1)/p)\} + \lambda(p) + \lambda(1-p)$
$\qquad = \mu(p)\{\lambda((p+q-1)/q) + \lambda((1-p)/q)\} + \lambda(q) + \lambda(1-q)$, for $p,q \in \,]0,1]$ with $p+q \geq 1$.

The proof of lemma *1.2.7* is based on a result due to Ng(1973).

LEMMA 1.2.8: Let $\emptyset = [a,b] \times [0,1]$ be a function satisfying the

Entropies of Order α

properties (b_1) to (b_5) of lemma 1.2.6 then $\Psi_{4,2}(u,v,p)$ is given by

(1.2.19) $\Psi_{4,2}(u,v,p) = p^\alpha u + (1-p)^\alpha v + h\{p^\alpha + (1-p)^\alpha - 1\}$, if $\alpha \neq 1$

(1.2.20) $\Psi_{4,2}(u,v,p) = pu + (1-p)v - h\{p \log p + (1-p)\log(1-p)\}$, if $\alpha = 1$

for all $u,v \in [a,b]$, $p \in [0,1]$ and where $0\log 0 = 0$ and h and $\alpha > 0$ are constants.

Proof of theorem 1.2.2:   We have

(1.2.21) $K_{n+1}(p_1,p_2,\ldots,p_{n+1}) = K_n(p_1+p_2,p_3,\ldots,p_{n+1}) + K_3(p_1,p_2,p_3+\ldots+p_{n+1})$
$\qquad\qquad - K_2(p_1+p_2,p_3+p_4+\ldots+p_{n+1})$

for all $(p_1,\ldots,p_{n+1}) \in S_{n+1}$ and $n > 2$. Also

(1.2.22) $K_2(p,1-p) = \Psi_{4,2}(K_2(1,0), K_2(1,0),p) = \Psi_{4,2}(k,k,p)$ and

(1.2.23) $K_3(p_1,p_2,p_3) = \Psi_{4,2}[K_2(p_1/(p_1+p_2), p_2/(p_1+p_2)), k, p_1+p_2]$
$\qquad\qquad = \Psi_{4,2}[\Psi_{4,2}(k,k,p_1/(p_1+p_2)), k, p_1+p_2],$

for every $(p_1,p_2,p_3) \in S_3$ with $p_1 + p_2 > 0$. The function $\Psi_{4,2}$ appearing in (1.2.22) and (1.2.23) is given by lemma 1.2.8. Hence

(1.2.24) $K_3(p_1,p_2,p_3) = h(p_1^\alpha + p_2^\alpha + p_3^\alpha - 1) + K(p_1^\alpha + p_2^\alpha + p_3^\alpha)$, for $\alpha \neq 1$
$\qquad\qquad = - h(p_1 \log p_1 + p_2 \log p_2 + p_3 \log p_3) + k$, for $\alpha = 1$

Hence using (1.2.24) in (1.2.21) we get

(1.2.25) $K_n(p_1,p_2,\ldots,p_n) = h\ H_{n,\alpha}(p_1,\ldots,p_n) + k\ \sum_{i=1}^{n} p_i^\alpha$

where h and k are real constants, $\alpha > 0$ with $h+k(2^{1-\alpha} - 1) > 0$.

Now the use of the postulates $E6$ and $E7$ prove the theorem completely.

1.3   GENERALIZATIONS AND OTHER MEASURES

A number of generalizations of the various measures of entropy are given recently and they will be discussed in this section.

(a) GENERALIZED FUNCTIONAL EQUATIONS

In this section we will discuss the measurable solutions of

two functional equations useful in the characterization of the entropies. The following theorem is taken from the paper of Kannappan and Ng (1973).

THEOREM 1.3.1: The most general measurable solutions of

(1.3.1) $F(x) + (1-x) G\{y/(1-x)\} = H(y) + (1-y) K\{x/(1-y)\}$

where $G, K: [0,1] \to \mathbb{R}$, $F, H: [0,1[\to \mathbb{R}$ for $x, y \in [0,1[$ with $x+y \in [0,1]$ are given by

(1.3.2) $\begin{cases} F(x) = A\,f(x) + B_1 x + D \\ G(y) = A\,f(y) + B_2 y + B_1 - B_4 \\ H(x) = A\,f(x) + B_3 x + B_1 + B_2 - B_3 - B_4 + D \\ K(y) = A\,f(y) + B_4 y + B_3 - B_2 \end{cases}$

for $x \in [0,1[$ and $y \in [0,1]$, where f is given by (1.1.3) and A, B_1, B_2, B_3, B_4 and D are arbitrary constants.

Proof: The proof depends on the following lemmas which we will state without proofs.

LEMMA 1.3.1: The functions F, H and K satisfying (1.3.1) can be expressed as affine compositions of G given by

(1.3.3) $H(y) = G(y) + a_1 y + b_1$, $x \in [0,1[$

(1.3.4) $F(x) = G(1-x) + a_2 x + b_2$, $x \in \,]0,1[$ and

(1.3.5) $K(x) = G(1-x) + a_3 x + b_3$, $x \in \,]0,1[$

where a_1, a_2, a_3, b_1, b_2 and b_3 are constants such that G is a solution of

(1.3.6) $G(1-x) + (1-x) G\{y/(1-x)\} = G(y) + (1-y) G\{(1-x-y)/(1-y)\}$
                                        $+ ax + by + c$

for $x \in \,]0,1[$, $y \in [0,1[$ with $x+y \in \,]0,1[$, where a, b, c are constants. For $y=0$, the equation (1.3.6) gives

$\qquad -x\,G(0) = a x + c$

for all $x \in \,]0,1[$. Hence $c = 0$. Thus (1.3.6), on substituting

(1.3.7) $u(x) = G(x) - (a+b)x + a$

for $x \in [0,1[$ takes the following form:

$(1.3.8) \quad u(1-x) + (1-x) u\{y/(1-x)\} = u(y) + (1-y) u\{(1-x-y)/(1-y)\}$

for $x \in]0,1[$, $y \in [0,1[$ with $x+y \in]0,1[$.

LEMMA 1.3.2: If A, B, C are Lebesgue measurable subsets of $\mathcal{R}$ with finite measure, then $x \to \mu(A \cap (1-x)B \cap (1-xC))$ is continuous. (See Hewitt and Stromberg, 1965).

LEMMA 1.3.3: If u is measurable in $]0,1[$ and satisfies (1.3.8) for all $x, y \in]0,1[$ with $x+y \in]0,1[$, then u is locally bounded and hence locally integrable.

LEMMA 1.3.4: The general measurable solution of (1.3.8), $x, y \in]0,1[$ with $x+y \in]0,1[$, is given by

$(1.3.9) \quad u(x) = A f(x)$

where A is an arbitrary constant and f is the Shannon's entropy function given by (1.1.3).

From lemmas 1.3.1 and 1.3.4 and the equation (1.3.7), it is easy to see that F, G, H and K are of the form

$$F(x) = A f(x) + d_1 x + c_1, \quad G(x) = A f(x) + d_2 x + c_2$$
$$H(x) = A f(x) + d_3 x + c_3, \quad K(x) = A f(x) + d_4 x + c_4$$

for $x \in]0,1[$ where c's and d's are constants. A direct substitution of F, G, H, K into (1.3.1) gives (1.3.2) on $]0,1[$. An examination at the boundary points 0 and 1 reveals that F, G, H, K have the form (1.3.2) on the respective domains.

An interesting corollary to theorem 1.3.1 is given below.

COROLLARY 1.3.1: For $F = G = H = K$

$(1.3.10) \quad F(x) = A[-x \log x - (1-x) \log(1-x)] + Bx$

where A and B are constants.

Under certain conditions (1.3.10) reduces to Shannon function (1.1.2) which is very important in the axiomatic characterization of Shannon entropy (1.1.1).

Next we give the measurable solutions for the following functional equation,

$(1.3.11) \quad F(x) + (1-x)^\alpha G\{y/(1-x)\} = H(y) + (1-y)^\alpha K\{x/(1-y)\}$

for $x, y \in [0,1[$, $x+y \in [0,1]$, where $G, K: [0,1] \to \mathcal{R}$ and $F, H: [0,1[\to \mathcal{R}$

and $\alpha(\neq 1) > 0$.

The theorem given below is due to Kannappan and Rathie(1973c).

THEOREM 1.3.2: The most general measurable solutions of *(1.3.11)* are given by

$$(1.3.12) \begin{cases} f(x) = A\, f_\alpha(x) + d_1 x^\alpha - c_2(1-x)^\alpha + c_1 \\ g(y) = A\, f_\alpha(y) + d_2 y^\alpha + c_2 \\ h(x) = A\, f_\alpha(x) + d_2 x^\alpha - c_4(1-x)^\alpha + c_1 \\ k(y) = A\, f_\alpha(y) + d_1 y^\alpha + c_4 \end{cases}$$

for $x \in [0,1[$, $y \in [0,1]$, where f_α is given by *(1.2.5)* and A, c_1, c_2, c_4, d_1 and d_2 are arbitrary constants.

The following lemmas will be required in the proof of the theorem.

LEMMA 1.3.5: If F, G, H and K satisfy the functional equation *(1.3.11)* then

$(1.3.13)$ $\quad F(x) = G(1-x) + [K(1) - K(0)]\, x^\alpha - G(1)(1-x)^\alpha + F(0)$, $x \in\,]0,1[$,

$(1.3.14)$ $\quad H(y) = G(y) - K(0)(1-y)^\alpha + F(0)$, $y \in [0,1[$,

$(1.3.15)$ $\quad K(x) = G(1-x) + [K(1) - K(0)]\, x^\alpha + [G(0) - G(1)](1-x)^\alpha$
$\hspace{4em} + F(0) - H(0)$, $x \in\,]0,1[$ and

$(1.3.16)$ $\quad u(1-x) + (1-x)^\alpha\, u[y/(1-x)] = u(y) + (1-y)^\alpha\, u[(1-x-y)/(1-y)]$

for $x \in\,]0,1[$, $y \in [0,1[$ with $x+y \in\,]0,1[$, where

$(1.3.17)$ $\quad u(x) = G(x) + [G(0) - G(1)]\, x^\alpha - G(0)$

Proof: For $x = 0$, *(1.3.11)* gives *(1.3.14)*. Taking $y=1-x$ in *(1.3.11)* and utilizing *(1.3.14)* one gets *(1.3.13)*. For $y=0$ in *(1.3.11)* and using *(1.3.13)*, we have *(1.3.15)*. Putting the expressions for F, H and K from *(1.3.13), (1.3.14)* and *(1.3.15)* respectively in *(1.3.11)*, we get

$(1.3.18)$ $\quad G(1-x) + (1-x)^\alpha\, G[y/(1-x)] = G(y) + (1-y)^\alpha\, G[(1-x-y)/(1-y)]$
$\hspace{4em} + [G(0) - G(1)](1-x-y)^\alpha + [F(0) - H(0) - K(0)](1-y)^\alpha$
$\hspace{4em} + G(1)(1-x)^\alpha$, $x \in\,]0,1[$, $y \in [0,1[$ with $x+y \in\,]0,1[$.

For $y=0$ the equation *(1.3.18)* gives $G(0) - H(0) + F(0) - K(0) = 0$. Hence *(1.3.18)* can be rewritten as

$(1.3.19)$ $\quad G(1-x) + (1-x)^\alpha\, G[y/(1-x)] = G(y) + (1-y)^\alpha\, G[(1-x-y)/(1-y)]$
$\hspace{4em} + G(1)(1-x)^\alpha - G(0)(1-y)^\alpha + [G(0) - G(1)](1-x-y)^\alpha$

for $x \in]0,1[$, $y \in [0,1[$ with $x+y \in]0,1[$. The equation (1.3.19) due to (1.3.17) gives (1.3.16) thus proving the lemma 1.3.5.

LEMMA 1.3.6: If u is measurable in $]0,1[$ and satisfies (1.3.16) for all $x,y \in]0,1[$ with $x+y \in]0,1[$, then u is locally integrable.

The proof of the above lemma is analogous to that of lemma 3 in Kannappan and Ng (1973).

LEMMA 1.3.7: The most general measurable solution of (1.3.16) for $x,y \in]0,1[$ with $x+y \in]0,1[$, is given by

(1.3.20) $\quad u(x) = A f_\alpha(x)$, $x \in]0,1[$

where A is an arbitrary constant and $f_\alpha(x)$ is defined in (1.2.5).

Proof: Following Kannappan and Ng (1973) it is easy to prove that u is differentiable infinitely many times in $]0,1[$. Now differentiating (1.3.16) first with respect to x and then the resulting expression with respect to y and then replacing $y/(1-x)$ by t and $x/(1-y)$ by $1-s$, we get

(1.3.21) $\quad (1-t)^{2-\alpha}[t u''(t) - (\alpha-1)u'(t)] = s^{2-\alpha}[(1-s)u''(s) + (\alpha-1)u'(s)]$

for $t,s \in]0,1[$. This gives

(1.3.22) $\quad u(t) = [\lambda/\{\alpha(\alpha-1)\}](1-t)^\alpha + c_2 t^\alpha - c_1/\alpha$, $t \in]0,1[$,

where λ, c_1 and c_2 are constants. Now (1.3.16) and (1.3.22) give

$$\lambda/[\alpha(\alpha-1)] = c_1/\alpha = c_2 = A \text{ (say)}$$

This proves lemma 1.3.7.

Proof of theorem 1.3.2: From (1.3.13), (1.3.14), (1.3.15), (1.3.17) and (1.3.20) it follows that

(1.3.23) $\quad \begin{cases} F(x) = A f_\alpha(x) + d_1 x^\alpha + c_1 + b_1(1-x)^\alpha \\ G(x) = A f_\alpha(x) + d_2 x^\alpha + c_2 \\ H(x) = A f_\alpha(x) + d_2 x^\alpha + c_1 + b_3(1-x)^\alpha \\ K(x) = A f_\alpha(x) + d_1 x^\alpha + c_4 + b_4(1-x)^\alpha \end{cases}$

for $x \in]0,1[$ where $A, d_1, d_2, c_1, c_2, c_4, b_1, b_3$ and b_4 are arbitrary constants and $f_\alpha(x)$ is given by (1.2.5). Direct substitution of (1.3.23) in (1.3.11) gives $b_1 = -c_2$, $b_3 = -c_4$ and $b_4 = 0$ giving the expressions in the theorem 1.3.2. An examination at the boundary reveals that F, G, H, K have the form (1.3.12) on the respective domains. This completes the proof of theorem 1.3.2.

When F, G, H and K are the same, (1.3.11) reduces to (1.2.6) with f_α replaced by F, the measurable solution of which is given in the following corollary to theorem 1.3.2.

COROLLARY 1.3.2: The most general measurable solution of (1.2.6) is

(1.3.24) $F(x) = A f_\alpha(x) + B x^\alpha$, $x \in [0,1]$

where A and B are arbitrary constants and $f_\alpha(x)$ is given by (1.2.5).

Note: For $F(0) = F(1)$ and $F(\tfrac{1}{2}) = 1$, (1.3.24) reduces to $f_\alpha(x)$, the non-additive entropy function of order α.

(b) OTHER GENERALIZATIONS

The most general expressions which include the various entropies are given in Rathie (1970,1971b,1973) and they are defined as follows.

DEFINITION 1.3.1

(1.3.25) $H_{n,\alpha}^{\beta_i}(P) = (\sum_{i=1}^{n} p_i^{\alpha+\beta_i-1} / \sum_{i=1}^{n} p_i^{\beta_i} -1)/(2^{1-\alpha} -1)$, $\alpha \neq 1$

which reduces to

(a_1) $H_{n,\alpha}^{\beta}(P)$ of Rathie (1971b) for all $\beta_i = \beta$, $i = 1,\ldots,n$;

(a_2) (1.2.2) when $\beta_i = 1$ for all $i = 1,\ldots,n$;

(a_3) (1.1.1) when $\beta_i = 1$ for all $i = 1,\ldots,n$ and $\alpha \to 1$.

DEFINITION 1.3.2

(1.3.26) $H_{n,1}^{\beta_i}(P) = - \sum_{i=1}^{n} p_i^{\beta_i} \log p_i / \sum_{i=1}^{n} p_i^{\beta_i}$

DEFINITION 1.3.3

(1.3.27) $\hat{H}_{n,\alpha}^{\beta_i}(P) = (1-\alpha)^{-1} \log[\sum_{i=1}^{n} p_i^{\alpha+\beta_i-1} / \sum_{i=1}^{n} p_i^{\beta_i}]$, $\alpha \neq 1$

Both (1.3.26) and (1.3.27) reduce to $H_{n,1}^{\beta}$ and $\hat{H}_{n,\alpha}^{\beta}$ respectively, of Kapur (1967) or Aczél and Daróczy (1963) when $\beta_i = \beta$ for all $i = 1,\ldots,n$.

When $\beta_i=1$ for all $i=1,\ldots,n$ in (1.3.26) it reduces to H_n given by (1.1.1). When $\beta_i=1$ for all $i=1,\ldots,n$ in (1.3.27) it yields $\hat{H}_{n,\alpha}$, that is (1.2.1). The expressions given by Varma (1966) are obtained from $\hat{H}_{n,\alpha}$ on taking $\alpha=\beta-n+1$ and $\alpha=\beta/n$ respectively.

Generalizations and Other Measures

Another generalization of *(1.1.1)* as given by Rathie (1971c) is as follows:

DEFINITION 1.3.4

(1.3.28) $\quad H_n(p_1,\ldots,p_n;\beta) = - \sum_{i=1}^{n} p_i^{\beta+1} \log p_i$

Some results for the expressions in *(1.3.25)*, *(1.3.26)*, *(1.3.27)* and *(1.3.28)* along with some of their particular cases will be included in Chapter 4 for $\sum_{i=1}^{n} p_i \leq 1$.

Various properties of the functions given in this section may be found in the papers mentioned above.

DEFINITION 1.3.5: Let $E_1, E_2, \ldots, E_n$ be a finite set of events having the probabilities $p_1, p_2, \ldots, p_n$ with $p_i \geq 0$, $\Sigma p_i = 1$ and the utilities $u_1, u_2, \ldots, u_n$ with $u_i \geq 0$. Then Belis and Guiasu (1968) defined the following generalization to Shannon entropy.

(1.3.29) $\quad I(u_1, u_2, \ldots, u_n; p_1, \ldots, p_n) = - K \Sigma u_i p_i \log p_i$

where K is an arbitrary positive constant. Clearly, for $u_1 = \ldots = u_n = 1$, and $K=1$, *(1.3.29)* reduces to Shannon entropy *(1.1.1)*. For further details, see the paper by Belis and Guiasu (1968).

DEFINITION 1.3.6: An information measure due to Hartley (1928) is defined by

(1.3.30) $\quad H_n(1/n, 1/n, \ldots, 1/n) = \log n$

and it is clearly a special case of Shannon entropy *(1.1.1)*.

DEFINITION 1.3.7: The index of diversity due to Gini (1912) is defined by

(1.3.31) $\quad D(p_1,\ldots,p_n) = 1 - \sum_{i=1}^{n} p_i^2$

Clearly *(1.3.31)* is related to *(1.2.2)* with $\alpha=2$.

1.4 CONTINUOUS ANALOGUES

Shannon's entropy for a continuous distribution with the density function $p(x)$ is defined by

(1.4.1) $\quad H = - \int_{-\infty}^{\infty} p(x) \log p(x) \, dx$

The entropy defined by *(1.4.1)* has most of the properties of that of the discrete case. The main difference between discrete and continuous cases is that the entropy measures the randomness of the chance variable in a well-defined way in the discrete case whereas the measurement is relative to the co-ordinate system in the continuous case.

In a similar way corresponding continuous analogues to (1.2.1), (1.2.2) etc., can be easily defined and their properties studied. Detailed discussion of Shannon's entropy of a continuous distribution can be seen from Kullback (1959), Ash (1965), Gallager (1968) and Dutta (1966). Some results regarding the dimension and entropy may be seen in Csiszár (1961,1962).

1.5 APPLICATIONS

Some of the quantities defined in this chapter have some interesting applications which can be seen from the following discussions. However it will be useful to investigate the various interpretations of these concepts in different disciplines and their applications.

Applications of the concept of entropy in Physics are discussed in Brillouin (1956). A collection of some papers in which possible applications of entropy in Chemistry, Biology and Psychology have been edited by Quastler (1953,1956).

(a) QUESTIONNAIRE THEORY

As an application of Shannon's entropy, we consider the problem of locating a square on a checkerboard. Since a checkerboard has 64 squares, 6 ($= \log_2 64$) questions will be sufficient to locate the square. These questions might take the following forms:

(Q_1) Is it one of the thirty two on the right half of the board? (Yes)

(Q_2) Is it one of the sixteen in the lower half of these 32 ? (Yes)

(Q_3) Is it one of the eight in the right half of these sixteen remaining ? (No)

(Q_4) Is it one of the four in the lower half of the eight remaining ? (No)

(Q_5) Is it one of the two in the right half of the four remaining? (Yes)

(Q_6) Is it the lower one of the two remaining ? (No)

For this and other similar problems, see Bending (1953).

Applications

One may have more such applications if one of the players announces "I am thinking of something ..." and others proceed to ask him questions to which he can answer "Yes" or "No". For more applications of this type see Picard (1965).

(b) CODING THEORY

One interpretation for *(1.1.1)* may be given as follows. The quantity $-\log p_i$, in Communication Theory, is usually known as the *information content* in the event E_i with probability p_i, or the amount of *self-information* associated with the event E_i. Thus *(1.1.1)* is an expected value.

Another interesting and important application of entropy *(1.1.1)* in Coding Theory is contained in the following discussion of Shannon's noiseless coding theorem.

Let $p_1,\ldots,p_n$ be the probabilities of n input symbols $x_1,\ldots,x_n$. Let x_i be represented by a sequence of n_i characters from the binary alphabet. Thus the *cost-function* is defined by

(1.5.1) $C = \Sigma\, p_i\, n_i$

The Shannon's noiseless coding theorem states that the minimum of $\Sigma p_i n_i$ is the entropy H_n, that is,

(1.5.2) $\Sigma p_i n_i \geq H_n(p_1,\ldots,p_n)$

with equality if and only if $n_i = -\log p_i$ for all i. For more details, see Ash (1965), Gallager (1968) or other books on Information Theory or Communication Theory.

Now we discuss a theorem due to Campbell (1965) and it can be taken as an application of the entropy of order α to Coding Theory.

Let $p_1,\ldots,p_n$ be the probabilities of n input symbols $x_1,\ldots,x_n$ which we wish to encode. We assume that $p_i > 0$ for $i = 1,\ldots,n$ and that $\Sigma p_i = 1$. Suppose that there is an alphabet of D symbols into which the input symbols are to be encoded. Let x_i be represented by a sequence of n_i characters from the alphabet. It is well-known (Feinstein, 1958) that there is a uniquely decipherable code with lengths $n_1,\ldots,n_n$ if and only if

(1.5.3) $\sum_{i=1}^{n} D^{-n_i} \leq 1$

For the situations where the cost is more nearly an exponential function of n_i, Campbell (1965) defined a code length of order t by

(1.5.4) $\quad L(t) = t^{-1} \log_D(\Sigma_{i=1}^n p_i D^{tn_i})$, $\quad 0 < t < \infty$

Clearly (1.5.4) is a generalization of (1.5.1) which is obtained by taking $t \to 0$ in (1.5.4). Now we will prove the following theorem.

THEOREM 1.5.1: Let $n_1, \ldots, n_n$ satisfy (1.5.3). Then

(1.5.5) $\quad (\log_2 D) L(t) \geq \hat{H}_{n,\alpha}(p_1, \ldots, p_n)$

where $\alpha = (1+t)^{-1}$, $0 < t < \infty$.

Proof: By Hölder's inequality

(1.5.6) $\quad (\Sigma_{i=1}^n x_i^p)^{1/p} (\Sigma_{i=1}^n y_i^q)^{1/q} \leq \Sigma_{i=1}^n x_i y_i$

where $p^{-1} + q^{-1} = 1$ and $p < 1$. In (1.5.6), let $p = -t$, $q = 1-\alpha$, $x_i = p_i^{-1/t} D^{-n_i}$ and $y_i = p_i^{1/t}$. The equation $p^{-1} + q^{-1} = 1$ implies that $\alpha = (1+t)^{-1}$. With these substitutions (1.5.6) becomes

$$(\Sigma p_i D^{tn_i})^{-1/t} (\Sigma p_i^\alpha)^{1/(1-\alpha)} \leq \Sigma D^{-n_i}$$

Therefore

(1.5.7) $\quad (\Sigma p_i D^{tn_i})^{1/t} \geq [(\Sigma p_i^\alpha)^{1/(1-\alpha)}]/\Sigma D^{-n_i} \geq (\Sigma p_i^\alpha)^{1/(1-\alpha)}$

where the last inequality follows from the assumption that (1.5.3) is satisfied. If we take logarithms to the base D of the first and third members of (1.5.7), we have the statement of theorem 1.5.1.

An easy calculation shows that we have equality in (1.5.5), (1.5.7) and (1.5.3) if

$$D^{-n_i} = p_i^\alpha / \Sigma p_i^\alpha$$

or if

(1.5.8) $\quad n_i = -\alpha \log_D p_i + \log_D(\Sigma_{j=1}^n p_j^\alpha)$

Thus, if we ignore the additional constraint that each n_i should be an integer, it is seen that the minimum possible value of $L(t)$ is $\hat{H}_{n,\alpha}$.

Further discussions and another coding theorem can be found in Campbell (1965). For further generalizations of this theorem, see Rathie (1970, 1972).

Jelinek (1968) developed and analysed a scheme for variable length coding of discrete memoryless fixed-rate sources in which buffer overflows result in code-word erasures at locations that are clearly specified to the user. The corresponding bounds on the probability of buffer overflow provide a linkup between source coding and generalized source entropy of order α. For further details, see Jelinek (1968).

(c) DIVERSITY IN HUMAN ECOLOGY

Some of the properties of Gini's diversity measure *(1.3.31)* were studied by Bhargava and Uppuluri (1971). For a geometric study of diversity, see Bhargava and Doyle (1974).

(d) ECONOMICS

Shannon's entropy may be given an interpretation as a measure of *income inequality* among a set of individuals. Applications of this to income distribution and stock market may be seen from Theil (1956, p.91-) and Cozzolino and Zahner (1973) respectively.

Shannon's entropy may be interpreted as an *inverse measure of industrial concentration* in allocation problems involving quantities such as market shares, outputs, number of employees per firm etc. For details, see Theil (1956, pp.290-292) and Horowitz (1970).

(e) MUSIC

Applications of Shannon's entropy to music can be seen in Siromoney and Rajagopalan (1964).

(f) LANGUAGES

Shannon's entropy has been extensively used in the analysis of the structure of languages. Various applications can be seen from Hyvärinen (1970), Siromoney (1963,1964) and Balasubrahmanyam and Siromoney (1968).

(g) PSYCHOLOGY

Information Theory is used in the areas of *positive and negative instances to concept formation, probabilistic functionalism, theories of behavior, theory of vision, perceptual grouping* and man's ability to transmit information. One of the greatest fascinations of Information Theory for the psychologists is that it offers a methodology for quantifying various concepts. For example, in order to describe *stimuli*, measures such as *topological information content* are used. Consider a group of points which are connected in certain ways. We may group the points into classes according to their topological properties, points within a class being topologically indistinguishable from one another. A measure $-\Sigma\, p\, \log p$ is usually applied with p interpreted as the proportion of points in each class.

The guessing-game technique is successfully used in the study of human organism's characteristics as an information channel. For a somewhat detailed description of the various specific examples the reader is referred to Attneave (1959).

1.6 OPEN PROBLEMS

Several authors have given characterization theorems for Shannon's entropy and its various generalizations as can be seen from the exercises given at the end of this chapter. The authors feel that many more interesting characterization theorems for the various entropies can be proved by other suitable choices of postulates. The following are some problems worth looking into.

1.1 Determine the various sets of postulates characterizing the quantities defined in *(1.2.1)*, *(1.3.25)*, *(1.3.26)*, *(1.3.27)* and *(1.3.28)* respectively.

1.2 Find the general solution of the functional equation
$f(x) + G(x)\, g(y/(1-x)) = h(y) + H(y)\, k(x/(1-y))$, $x, y \in [0,1[$, $x+y \in [0,1]$.
Also discuss its special cases.

1.3 Discuss the various properties of the information measure defined in *(1.3.29)* and find the various sets of postulates characterizing this measure.

EXERCISES

1.1 Show that the function f satisfying the equation *(1.1.13)* and which is
  (a) (Daróczy, 1969), continuous in $[0,1]$ or continuous at $x = 0$ or
  (b) (Tverberg, 1958), Lebesgue-integrable in $[0,1]$ (see theorem *1.1.2*) or
  (c) (Kendall, 1964), monotone non-decreasing in $[0,\frac{1}{2}]$ or
  (d) (Lee, 1964), Lebesgue-measurable in $]0,1[$ or
  (e) (Daróczy and Katai, 1970), non-negative bounded in $[0,1]$
  is uniquely determined as *(1.1.2)*.

1.2 (Khinchin, 1953). Prove that the following postulates determine the function $H_n(p_1,\ldots,p_n)$ uniquely up to a multiplicative constant:
  (a) Inequality: *(1.1.11)*

Exercises

(b) Symmetry: *(1.1.11)*.
(c) Continuity: Property *(vii)* of §*1.1*.
(d) Expansibility: *(1.1.6)*.
(e) Strong additivity: *(1.1.10)*.

1.3 (Fadeev, 1956, Rényi, 1959,1961). Show that the following three conditions determine the function $H_n(p_1,\ldots,p_n)$ uniquely up to a multiplicative constant:
(a) $H_2(p,1-p)$ is a continuous function of $p \in [0,1]$.
(b) Symmetry: *(1.1.7)*.
(c) Recursivity: *(1.1.8)*.

1.4 (Kendall, 1964). Let S_k denote the set of k-ples of real numbers $(p_1,\ldots,p_k)$ such that $p_j \geq 0$, $j=1,\ldots,k$; $p_1+\ldots+p_k=1$. Let $H_k(p_1,\ldots,p_k)$ be the entropy on S_k.
(a) Let H_2 and H_3 be real symmetric functions on the interior of S_2 and S_3, respectively.
(b) Let the function h defined by $h(t) = H_2(t,1-t)$, $0 < t < 1$, be non-decreasing for $0 < t \leq \frac{1}{2}$ and let $h(\frac{1}{2}) = 1$.
(c) Let H_2 and H_3 be connected by the relation $H_3[tp_1,(1-t)p_1,p_2] = H_2(p_1,p_2) + p_1 H_2(t,1-t)$, when (p_1,p_2) lies in the interior of S_2 and $0 < t < 1$. Then show that H_2 and H_3 are given by the relation $H_k(p_1,\ldots,p_k) = -\sum_{j=1}^{k} p_j \log p_j$ on the interior of S_2 and S_3 respectively.

1.5 (Lee, 1964). Show that $H_k(p_1,\ldots,p_k)$ where $p_i > 0$, $i = 1,\ldots,k$, $\sum_{i=1}^{k} p_i = 1$ is uniquely determined as *(1.1.1)* under the following assumptions.
(a) H_k is permutation-symmetric for $k = 2,3$
(b) Define $h(t) = H_2(t,1-t)$, $0 < t < 1$. Let $h(t)$, be a finite real-valued Lebesgue measurable function defined on $]0,1[$ with $h(\frac{1}{2}) = 1$.
(c) For $0 < t < 1$, $k > 1$ and $p_1,\ldots,p_k > 0$, $\sum_{j=1}^{k} p_j = 1$, let $H_{k+1}[tp_1,(1-t)p_1,p_2,\ldots,p_k] = H_k(p_1,\ldots,p_k) + p_1 H_2(t,1-t)$.

1.6 (Pintacuda, 1966). Show that $H_n(p_1,\ldots,p_n)$, $p_i \geq 0$, $i = 1,\ldots,n$, $\Sigma p_i = 1$ is uniquely determined as *(1.1.1)* by the following set of postulates.
(a) $H_n(p_1,\ldots,p_n)$ is a continuous symmetric function.
(b) $H_{n+1}(p_1,\ldots,p_n,0) = H_n(p_1,\ldots,p_n)$.
(c) $H_{n+1}(p_1,\ldots,p_{n-1},q_1,q_2) - H_n(p_1,\ldots,p_{n-1},p_n)$, where $p_n = q_1+q_2$, should not depend upon the variables $p_1,\ldots,p_{n-1}$.
(d) $H_{mn}(p_1 q_1,\ldots,p_n q_1,\ldots,p_1 q_m,\ldots,p_n q_m) = H_n(p_1,\ldots,p_n)$

$$+ H_m(q_1,\ldots,q_m).$$
(e) $H_2(\frac{1}{2},\frac{1}{2}) = 1$.

1.7 (Borges, 1967). Let $H_n(p_1,\ldots,p_n)$, $p_i > 0$, $\Sigma p_i = 1$ satisfy the postulates,

(a) $H_n(p_1,\ldots,p_n) = H_{n-1}(p_1+p_2,p_3,\ldots,p_n) + (p_1+p_2)H_2(p_1/(p_1+p_2),$
$$p_2/(p_1+p_2)) \text{ for } n \geq 3,$$

(b) $H_3(p_1,p_2,p_3)$ for $p_1+p_2+p_3 = 1$ is symmetric in p_1,p_2,p_3,

(c) $H_2(\frac{1}{2},\frac{1}{2}) = 1$,

(d) $H_2(p,1-p)$ is monotonic non-decreasing in $0 < p \leq \frac{1}{2}$,

then show that H_n is Shannon's entropy (1.1.1) for $n \geq 2$.

1.8 (Daróczy, 1967). Let $H_n(p_1,\ldots,p_n)$, $p_i \geq 0$, $\Sigma_{i=1}^n p_i = 1, n = 2, 3,$ ... satisfy the postulates,

(a) $H_n(p_1,\ldots,p_n)$ is symmetric in $p_1,\ldots,p_n$,

(b) $H_n(p_1,\ldots,p_n) - H_{n-1}(p_1+p_2,p_3,\ldots,p_n) = \Delta_{n-1}(p_1,p_2), n \geq 3$, where $\Delta_2(p_1,p_2)$ is a continuous function in $\{(p_1,p_2) | p_1 \geq 0, p_2 \geq 0, p_1+p_2 \leq 1\}$,

(c) $H_{n+1}(p_1,\ldots,p_n,0) = H_n(p_1,\ldots,p_n)$,

(d) $H_{2n}[p_1 q, p_1(1-q),\ldots,p_n q, p_n(1-q)] = H_n(p_1,\ldots,p_n)$
$$+ H_2(q,1-q) \text{ for all } q \in [0,1],$$

(e) $H_2(\frac{1}{2},\frac{1}{2}) = 1$,

then show that H_n is Shannon's entropy given in (1.1.1).

1.9 (Daróczy, 1967). Show that Shannon's entropy (1.1.1) is uniquely obtained from the following postulates.

(a) $H_n(p_1,\ldots,p_n) = \Sigma_{i=1}^n f(p_i)$ where f is a continuous function of $p_i \geq 0$, $\Sigma p_i = 1$.

(b) $H_2(\frac{1}{2},\frac{1}{2}) = 1$.

(c) $H_{2n}[p_1 q, p_1(1-q),\ldots,p_n q, p_n(1-q)] = H_n(p_1,\ldots,p_n)$
$$+ H_2(q,1-q) \text{ for all } q \in [0,1].$$

1.10 (Forte and Daróczy, 1968). Let $H_n(p_1,\ldots,p_n)$ be a mapping of the set $\Delta_n = \{(p_1,\ldots,p_n) : p_i \geq 0, \Sigma_{i=1}^n p_i = 1\}$ into the set of real numbers $\mathcal{R}$ for $n = 2, 3, \ldots$. Let

(a) $H_n(p_1,\ldots,p_n)$ be a symmetric function of its arguments for $n = 4$;

(b) $H_n(p_1,\ldots,p_n) - H_{n-1}(p_1+p_2,p_3,\ldots,p_n) = \emptyset(p_1,p_2)$, $n = 3, 4, \ldots$

(c) $H_{2n}[p_1 q, p_1(1-q),\ldots,p_n q, p_n(1-q)] = H_n(p_1,\ldots,p_n)$
$$+ H_2(q,1-q), \text{ where } q \in [0,1] \text{ for } n=2 \text{ and } 3;$$

(d) $H_2(\frac{1}{2},\frac{1}{2}) = 1$;

(e) $\emptyset(p_1,p_2) > k$ for every $(p_1,p_2) \in \{(x,y) : x \geq 0, y \geq 0, x+y \leq 1\}$ where k is a finite constant.

Exercises 31

Then

(i) $H_2(p,1-p)$ is Lebesgue measurable in $]0,1[$ and

(ii) $\lim_{p \downarrow 0} H_2(p,1-p) = 0$, are equivalent and that each one of them implies $H_n(p_1,\ldots,p_n) = -\sum_{i=1}^{n} p_i \log p_i$, for all $(p_1,\ldots,p_n) \in \Delta_n$ for $n = 2,3,\ldots$

1.11 (Daróczy, 1969). If f is a solution of the functional equation (1.1.13) and the entropy of a distribution is defined by (1.1.4) then derive (1.1.6), (1.1.7), (1.1.8) and (1.1.10).

1.12 (Daróczy-Katai, 1970). If $H_n(p_1,\ldots,p_n)$, for $p_i \geq 0$, $\sum_{i=1}^{n} p_i = 1$, for $n = 2,3,\ldots$ satisfy the following postulates:
(a) $H_3(p_1,p_2,p_3)$ is symmetric,
(b) $0 \leq H_2(p,1-p) \leq H_2(\frac{1}{2},\frac{1}{2}) = 1$ for all $p \in [0,1]$,
(c) $H_n(p_1,\ldots,p_n) - H_{n-1}(p_1+p_2,p_3,\ldots,p_n) = (p_1+p_2)H_2(p_1/(p_1+p_2), p_2/(p_1+p_2))$, for $n \geq 3$, then $H_n(p_1,\ldots,p_n)$ is Shannon's entropy given by (1.1.1) for all $n = 2,3,\ldots$

1.13 (Aczél-Daróczy, 1963). For $(p_1,\ldots,p_n)$, $p_i > 0$, $\sum_{i=1}^{n} p_i = 1$ let the entropy be of the form $H_n(p_1,\ldots,p_n) = g^{-1}[\sum_{i=1}^{n} p_i g(-\log p_i)]$ where, (a) g is continuous in $[0,\infty[$ and (b) $\{x \, g(-\log x)$ for $x \in]0,1]$, 0 for $x = 0\}$ is strictly a convex function in $[0,1]$. Also let H_n satisfy (1.1.9) for $p_i > 0$, $q_i > 0$, $\sum_{i=1}^{n} p_i = 1$, $\sum_{j=1}^{m} q_j = 1$. Then prove that either H_n is Shannon's entropy defined in (1.1.1) or the entropy of order α ($\alpha > 0, \alpha \neq 1$) defined in (1.2.1).

1.14 (Daróczy, 1964). Prove the above problem 1.13 by replacing (b) by (b_1) $[\lim_{x \to 0} x \, g(-\log x)] = 0$, and (b_2) g is monotonic increasing in $[0,\infty[$.

1.15 (Aczél-Daróczy, 1963). Prove the exercise 1.14 when (b_1) is replaced by (b_3) $\lim_{p \to 1} H_2(p,1-p) = 0$.

1.16 (Campbell, 1966). Let $\emptyset$ be a continuous strictly nonotonic increasing function which satisfies the additivity condition:
$\emptyset^{-1}(\sum_{i=1}^{m}\sum_{j=1}^{k} p_i q_j \emptyset(n_i+m_j)) = \emptyset^{-1}(\sum_{i=1}^{m} p_i \emptyset(n_i)) + \emptyset^{-1}(\sum_{j=1}^{k} q_j \emptyset(m_j))$
and let $H(p_o,\emptyset) = \log_D m$ when $p_o = (1/m,\ldots,1/m)$. Then the generalized entropy $H(P,\emptyset) = \inf_{N \in S} L(P,N,\emptyset)$, where S is the set of all real distributions $N = (n_1,\ldots,n_m)$ for which $\sum_{i=1}^{m} D^{-n_i} \leq 1$ is satisfied and $L(P,N,\emptyset) = \emptyset^{-1}(\sum_{i=1}^{m} p_i \emptyset(n_i))$, must be the entropy of order α, given by (1.2.1) for some $\alpha > 0$.

1.17 (Rathie-Kannappan, 1971). Find the most general solution f of the functional equation

$f(x) + g(x) f(y/(1-x)) = f(y) + g(y) f(x/(1-y))$, for $x,y \in [0,1[$ with $x+y \in [0,1]$ where $g: I \to \mathcal{R}$ is a given function satisfying the functional equation $g(x+y-xy) = g(x) g(y)$, for $x,y \in [0,1]$ and $f: I \to \mathcal{R}$, satisfying further $f(0) = f(1)$ and $f(\frac{1}{2}) = 1$.

1.18 (Havrda-Charvát, 1967). Show that Shannon's entropy (1.1.1) and the non-additive entropy of order α, (1.2.2) are uniquely determined by the following four postulates on $H(p_1,\ldots,p_m;\alpha)$.
(a) $H(p_1,\ldots,p_n;\alpha)$ is continuous in the region $p_i \geq 0$, $\sum_{i=1}^{n} p_i = 1$, $\alpha > 0$.
(b) $H(\frac{1}{2},\frac{1}{2};\alpha) = 1$.
(c) $H(p_1,\ldots,p_{i-1},0,p_{i+1},\ldots,p_n;\alpha) = H(p_1,\ldots,p_{i-1},p_{i+1},\ldots,p_n;\alpha)$ for every $i = 1,2,\ldots,n$.
(d) $H(p_1,\ldots,p_{i-1},q_{i_1},q_{i_2},p_{i+1},\ldots,p_n;\alpha) = H(p_1,\ldots,p_{i-1},p_i,p_{i+1},\ldots,p_n;\alpha) + p_i^{\alpha} H(q_{i_1}/p_i, q_{i_2}/p_i;\alpha)$, for every $q_{i_1} + q_{i_2} = p_i > 0$, $i = 1,2,\ldots,n$, $\alpha > 0$.

1.19 (Daróczy, 1970). Let $H_{n,\alpha}: S_n \equiv \{(p_1,\ldots,p_n): p_i \geq 0, \sum_{i=1}^{n} p_i = 1\} \to \mathcal{R}$ ($n = 2,3,\ldots$) be a sequence of mappings and let α be a positive number different from unity. If $H_{n,\alpha}$ satisfies the following conditions:
(a) $H_{3,\alpha}(p_1,p_2,p_3)$ is a symmetric function of its variables;
(b) $H_{2,\alpha}(\frac{1}{2},\frac{1}{2}) = 1$;
(c) $H_{n,\alpha}(p_1,\ldots,p_n) - H_{n-1,\alpha}(p_1+p_2,p_3,\ldots,p_n)$
$= (p_1+p_2)^{\alpha} H_{2,\alpha}(p_1/(p_1+p_2), p_2/(p_1+p_2))$, for all $(p_1,\ldots,p_n) \in S_n$, $n = 3,4,\ldots$, $p_1+p_2 > 0$; then $H_{n,\alpha}$ is uniquely given by (1.2.2).

1.20 (Forte, 1973). Let S_n be the set of all probability distributions $(p_1,\ldots,p_n)$. Then prove that Shannon's entropy is uniquely determined by the following postulates: (a) Symmetry: (1.1.7) for every $n \geq 2$, (b) Expansibility: (1.1.6) for every $n \geq 2$, (c) Additivity: (1.1.9) for every $m \geq 2$, $n \geq 2$, (d) Sub-additivity: (1.1.12) for $m \geq 2$, $n \geq 2$, (e) Continuity at 0, that is, $\lim_{p \to 0^+} H_2(p,1-p) = H_2(0,1)$ and (f) Normalization: $H_2(\frac{1}{2},\frac{1}{2}) = 1$.

1.21 (Rathie, 1972). Let $P = (p_1,\ldots,p_m)$, $p_i \geq 0$, $\Sigma p_i = 1$, $N = (n_1,\ldots,n_m)$ satisfy the inequality $\Sigma p_i^{\beta-1} D^{-n_i} \leq \Sigma p_i^{\beta}$. For $\alpha = (1\ t)^{-1}$, $-1 \leq t \leq \infty$, $t \neq 0$ and $n_i = -\log_D p_i + \log_D (\Sigma p_j^{\alpha+\beta-1}/\Sigma p_j^{\beta})$, prove that $\hat{H}_{n,\alpha}^{\beta}(p_1,\ldots,p_n) = \inf [t^{-1} \log_D(\Sigma p_i^{\beta} D^{tn_i}/\Sigma p_i^{\beta})]$.

1.22 (Daróczy, 1971). Let f be a real valued function defined on $[0,1]$, satisfying the functional equation

$\sum_{i=1}^{2} \sum_{j=1}^{3} f(x_i y_j) = \sum_{i=1}^{2} f(x_i) + \sum_{j=1}^{3} f(y_j)$ for $x_i \geq 0$, $y_j \geq 0$, $x_1+x_2 = 1$, $y_1+y_2+y_3 = 1$ and the condition $f(1) = 0$. If f is measurable on $]0,1[$, then show that $f(x) = c\, x \log x$, for $x \in [0,1]$, where c is a constant.

1.23 Show that the maximum of $H = -\int p(x) \log p(x)\, dx$ subject to $\int x^2 p(x)\, dx = \sigma^2$ and $\int p(x)\, dx = 1$ is attained for $p(x) = (2\pi)^{-\frac{1}{2}} \sigma^{-1} \exp[-x^2/(2\sigma^2)]$ and that the maximum value of H is $\log_e((2\pi e)^{\frac{1}{2}} \sigma)$.

1.24 If x is limited to be non-negative (that is, $p(x) = 0$ for $x < 0$) and the first moment of x is fixed at a (that is, $a = \int_0^\infty x\, p(x)\, dx$) then prove that the maximum of $H = -\int p(x) \log p(x)\, dx$ occurs when $p(x) = (1/a) \exp(-x/a)$ and is equal to $\log_e(ea)$.

1.25 Use Stirling's formula to show that for large n_i, $i = 1,\ldots,k$ $\log N$ is approximately equal to $-n \sum_{i=1}^{k} p_i \log p_i$ where $p_i = n_i/n$, $n = n_1 + \ldots + n_k$ and $N = n!/(n_1!\, n_2! \ldots n_k!)$.

CHAPTER 2

THE CONCEPT OF DIRECTED DIVERGENCE

2.0 INTRODUCTION

This chapter deals with the concept of directed divergence between two discrete distributions, some of their generalizations and some measures involving more than two distributions. Different definitions and some of the properties are pointed out and characterizations are also given. This measure is widely used in Information Theory, Communication Theory, forecasting, statistical Inference, Statistical Mechanics and other disciplines.

All the theorems discussed in this chapter are taken from the articles, Kannappan and Rathie (1970,1972,1973,1973a,1973b,1973c,1974), Rathie (1970,1971,1973) and Rathie and Kannappan (1972,1973).

All the problems listed in the excercises of this chapter are taken from Campbell (1972), Kannappan (1972,1972a,1973), Kannappan and Rathie (1970,1973,1973b), Kannappan and Ng (1973,1974), Ng (1974) and Rathie (1973).

Some of the specific applications and interpretations are discussed and some open problems are also given.

2.1 DIRECTED DIVERGENCE BETWEEN TWO DISTRIBUTIONS

Let $P = (p_1,\ldots,p_n)$, $p_i \geq 0$, $\Sigma p_i = 1$ and $Q = (q_1,\ldots,q_n)$, $q_i \geq 0$, $\Sigma q_i = 1$ be two finite discrete probability distributions. In other words $P \in S_n$ and $Q \in S_n$. Then a measure of directed divergence between P and Q is defined as follows.

(a) DEFINITION 2.1.1

(2.1.1) $\quad I_n(p_1,\ldots,p_n;q_1,\ldots,q_n) = \Sigma p_i \log (p_i/q_i)$

In *(2.1.1)* and throughout this section the following convention will be followed unless otherwise specified: $0 \log 0 = 0$, if $q_i = 0$, the corresponding $p_i = 0$ and $p_i \log(p_i/q_i) = p_i(\log p_i - \log q_i)$. The representation $0 \log 0 = 0$ is not used in the proofs of characterization theorems which will be discussed later. An interesting special case of *(2.1.1)* for $n = 2$ is

(2.1.2) $\quad I_2(p, 1-p : q, 1-q) = p \log(p/q) + (1-p) \log((1-p)/(1-q))$

where $(p,q) \in J \equiv]0,1[\times]0,1[\cup \{(0,y)\} \cup \{(1,y')\}$ with $y \in [0,1[$ and $y' \in]0,1]$. The quantity I_2 given in *(2.1.2)* is of fundamental importance and it will be seen from the characterization theorems discussed later on and therefore we will call $I_2(p, 1-p : q, 1-q)$ as the *directed divergence function*.

In the characterization theorems to be discussed later in this section some of the postulates are common to different theorems. In order to avoid repetitions we will list here some of the properties of *(2.1.1)* and *(2.1.2)* before discussing the theorems.

(b) PROPERTIES OF DIRECTED DIVERGENCE

These properties are classified into different categories and are listed below.

(i)   Non-negativity:

(2.1.3) $\quad I_n(P:Q) \geq 0$

with equality if and only if $p_i = q_i$ for all i.

(ii)   Expansibility:

(2.1.4) $\quad I_{n+1}(P, 0 : Q, 0) = I_n(P:Q)$

(iii) Symmetry: $I_n(P:Q)$ is symmetric in pairs $\{p_i, q_i\}$ for $i = 1, \ldots, n$. That is,

(2.1.5) $\quad I_n(p_{r_1}, \ldots, p_{r_n} : q_{r_1}, \ldots, q_{r_n}) = I_n(p_1, \ldots, p_n : q_1, \ldots, q_n)$

where $\{r_1, \ldots, r_n\}$ is an arbitrary permutation of $\{1, 2, \ldots, n\}$.

(iv) Recursivity:

(2.1.6) $\quad I_n(p_1, \ldots, p_n : q_1, \ldots, q_n) = I_{n-1}(p_1 + p_2, p_3, \ldots, p_n : q_1 + q_2, q_3, \ldots, q_n)$
$\quad\quad\quad + (p_1 + p_2) I_2(p_1/(p_1+p_2), p_2/(p_1+p_2) : q_1/(q_1+q_2), q_2/(q_1+q_2))$,

for $p_1 + p_2$, $q_1 + q_2 > 0$ and for all $n = 3, 4, \ldots$. This property suggests the way in which the measures are added up when the union of two mutually exclusive events are considered. In other words, if an event is split into two mutually exclusive events then the measure is a weighted sum with the weights being probabilities as given above.

(v) *Additivity:* For $p_i, q_j, r_i, s_j \geq 0$, $i=1,2,\ldots,m$; $j=1,2,\ldots,n$; $\sum_{i=1}^{m} p_i = \sum_{j=1}^{n} q_j = \sum_{i=1}^{m} r_i = \sum_{j=1}^{m} s_j = 1$,

(2.1.7) $\quad I_{mn}(p_1q_1,\ldots,p_1q_n,p_2q_1,\ldots,p_2q_n,\ldots,p_mq_1,\ldots,p_mq_n:r_1s_1,\ldots,r_1s_n,$

$r_2s_1,\ldots,r_ms_1,\ldots,r_ms_n) = I_m(p_1,\ldots,p_m:r_1,\ldots,r_m)$

$+ I_n(q_1,\ldots,q_n:s_1,\ldots,s_n)$

This property can be associated with the probabilities corresponding to mutually independent events.

(vi) *Strong Additivity:* For $\sum_{i=1}^{n} \sum_{j=1}^{m} p_{ij} = 1$, $p_j = \sum_{i=1}^{n} p_{ij} > 0$,

$q_j = \sum_{i=1}^{n} q_{ij} > 0$, $\sum_{i=1}^{n}\sum_{j=1}^{m} q_{ij} = 1$, one has

(2.1.8) $\quad I_{mn}(p_{11},\ldots,p_{nm}:q_{11},\ldots,q_{nm}) = I_m(p_1,\ldots,p_m:q_1,\ldots,q_m)$

$+ \sum_{j=1}^{m} p_j I_n(p_{1j}/p_j,\ldots,p_{nj}/p_j:q_{1j}/q_j,\ldots,q_{nj}/q_j)$.

(vii) *Continuity:* $I_n(P:Q)$ is a continuous function of its $2n$ variables.

(viii) *Functional Equation:* The equation (2.1.2) satisfies the functional equation

(2.1.9) $\quad f(x,y) + (1-x) f(u/(1-x), v/(1-y))$

$= f(u,v) + (1-u) f(x/(1-u), y/(1-v))$

for $x,y,u,v \in [0,1[$ with $x+u, y+v \in [0,1]$

$= f(x+u, y+v) + (x+u) f(x/(x+u), y/(y+v))$

for $x,y,u,v \in [0,1[$, $x+u, y+v \in]0,1]$.

(c) *CHARACTERIZATIONS*

This section deals mainly with two characterization theorems for directed divergence given in (2.1.1). The first theorem, due to Kannappan and Rathie (1973a), will be proved with the help of the following lemmas. Let the function $I_n(P:Q)$ satisfy the following postulates.

A1. *Recursivity:* (2.1.6)
A2. *Symmetry:* (2.1.5) for $n = 3$
A3. *Derivative:* Let $f(p,q) = I_2(p, 1-p:q, 1-q)$ for $(p,q) \in J$ have continuous first partial derivatives with respect to both variables $p,q \in]0,1[$

A4. *Normalization:* $I_2(2/3,1/3:1/3,2/3) = 1/3$
A5. *Nullity:* $I_2(p,1-p:p,1-p) = 0$ for $p \in\,]0,1[$

Then the following lemmas can be proved.

LEMMA 2.1.1: $I_2(p,1-p:q,1-q)$ is symmetric. That is, $f(p,q) = f(1-p,1-q)$, for $(p,q) \in J$.

Proof: For $n = 3$, postulate A1 takes the following form:

(2.1.10) $\quad I_3(p_1,p_2,p_3:q_1,q_2,q_3) = I_2(p_1+p_2,p_3:q_1+q_2,q_3)$

$\quad\quad\quad + (p_1+p_2)\, I_2(p_1/(p_1+p_2),p_2/(p_1+p_2):q_1/(q_1+q_2),q_2/(q_1+q_2))$

for $p_1+p_2,\, q_1+q_2 > 0$. Also for $n = 3$ the postulate A1 gives

(2.1.11) $\quad I_3(p_2,p_1,p_3:q_2,q_1,q_3) = I_2(p_2+p_1,p_3:q_2+q_1,q_3)$

$\quad\quad\quad + (p_2+p_1)\, I_2(p_2/(p_2+p_1),p_1/(p_2+p_1):q_2/(q_2+q_1),q_1/(q_2+q_1))$

for $p_2+p_1,\, q_2+q_1 > 0$. Hence (2.1.10), (2.1.11) and postulate A2 give

(2.1.12) $\quad I_2(p_1/(p_1+p_2),p_2/(p_1+p_2):q_1/(q_1+q_2),q_2/(q_1+q_2))$

$\quad\quad\quad = I_2(p_2/(p_1+p_2),p_1/(p_1+p_2):q_2/(q_1+q_2),q_1/(q_1+q_2))$

for $p_1+p_2,\, q_1+q_2 > 0$. This completes the proof. In particular, (2.1.12) yields,

(2.1.13) $\quad f(0,0) = f(1,1)$

LEMMA 2.1.2: f satisfies the pair of functional equations

(2.1.14) $\quad f(p_1+p_2,q_1+q_2) + (p_1+p_2)\, f(p_1/(p_1+p_2),q_1/(q_1+q_2))$

$\quad\quad\quad = f(p_1,q_1) + (1-p_1)\, f(p_2/(1-p_1),q_2/(1-q_1))$

$\quad\quad\quad = f(p_2,q_2) + (1-p_2)\, f(p_1/(1-p_2),q_1/(1-q_2))$

for $p_1,p_2,q_1,q_2 \in [0,1[\,,\; p_1+p_2,q_1+q_2 \in\,]0,1]$.

Proof: Postulate A2 implies

(2.1.15) $\quad I_3(p_1,p_2,p_3:q_1,q_2,q_3) = I_3(p_2,p_3,p_1:q_2,q_3,q_1)$

$\quad\quad\quad = I_3(p_3,p_1,p_2:q_3,q_1,q_2)$

Now from the postulate A1 the proof follows.

LEMMA 2.1.3: The function f for $x,y \in\,]0,1[$ is given by

(2.1.16) $\quad f(x,y) = a[x\, \log(x/2) + (1-x)\, \log\{(1-x)/2\}] + bx\, \log\{y/(1-y)\} + g(y)$

where a and b are constants and g is a function of y only.

Proof: Let the partial derivative of f with respect to the first variable be denoted by f_1. Now differentiating partially with respect to p_1 the first and third equation pair of *(2.1.14)* yield

(2.1.17) $\quad f_1(p_1+p_2, q_1+q_2) + f(p_1/(p_1+p_2), q_1/(q_1+q_2))$

$\qquad + \{p_2/(p_1+p_2)\} f_1(p_1/(p_1+p_2), q_1/(q_1+q_2)) = f_1(p_1/(1-p_2), q_1/(1-q_2))$

for $p_1, q_1 \in]0,1[$, $p_2, q_2 \in [0,1[$ with $p_1+p_2, q_1+q_2 \in]0,1]$. Also differentiating partially with respect to p_2 the first and second equation pair of *(2.1.14)* give

(2.1.18) $\quad f_1(p_1+p_2, q_1+q_2) + f(p_1/(p_1+p_2), q_1/(q_1+q_2))$

$\qquad - \{p_1/(p_1+p_2)\} f_1(p_1/(p_1+p_2), q_1/(q_1+q_2)) = f_1(p_2/(1-p_1), q_2/(1-q_1))$

for $p_2, q_2 \in]0,1[$, $p_1, q_1 \in [0,1[$ with $p_1+p_2, q_1+q_2 \in]0,1]$. Now subtracting *(2.1.18)* from *(2.1.17)*, one gets

(2.1.19) $\quad f_1(p_1/(p_1+p_2), q_1/(q_1+q_2)) = f_1(p_1/(1-p_2), q_1/(1-q_2))$

$\qquad - f_1(p_2/(1-p_1), q_2/(1-q_1))$

for $p_1, p_2, q_1, q_2 \in]0,1[$, $p_1+p_2, q_1+q_2 \in]0,1]$. Substituting $p_1 = xy/(1+y+xy)$, $p_2 = y/(1+y+xy)$, $q_1 = uv/(1+v+uv)$ and $q_2 = v/(1+v+uv)$ in *(2.1.19)*, one has

(2.1.20) $\quad f_1(x/(1+x), u/(1+u)) = f_1(xy/(1+xy), uv/(1+uv)) - f_1(y/(1+y), v/(1+v))$

for $x, y, u, v \in]0, \infty[$. Let

(2.1.21) $\quad F(x,u) = f_1(x/(1+x), u/(1+u))$, for $x, u \in]0, \infty[$.

Since f_1 is continuous F is also continuous. Now *(2.1.20)* with the help of *(2.1.21)* gives

(2.1.22) $\quad F(x,u) + F(y,v) = F(xy, uv)$, for $x, y, u, v \in]0, \infty[$.

Putting $u=v=1$ in *(2.1.22)*, one gets

(2.1.23) $\quad F(x,1) + F(y,1) = F(xy,1)$, for $x, y \in]0, \infty[$.

Equation *(2.1.23)* is the well-known Cauchy functional equation (Aczél(1966),p.215) of which the continuous solution is given by

(2.1.24) $\quad F(x,1) = a \log x$, for $x \in]0, \infty[$,

where a is an arbitrary real constant. Similarly taking $x=y=1$ in

$(2.1.22)$ it is seen that

$(2.1.25) \quad F(1,u) = b \log u, \quad u \in]0,\infty[$,

where b is an arbitrary real constant. Again taking $u=y=1$ in $(2.1.22)$ one gets

$(2.1.26) \quad F(x,v) = F(x,1) + F(1,v)$

Thus $(2.1.24)$, $(2.1.25)$ and $(2.1.26)$ give

$(2.1.27) \quad F(x,v) = a \log x + b \log v$, for $x,v \in]0,\infty[$.

Hence $(2.1.27)$ and $(2.1.21)$ give

$(2.1.28) \quad f_1(x,y) = a \log \{x/(1-x)\} + b \log \{y/(1-y)\}$, for $x,y \in]0,1[$

which on integration proves lemma $2.1.3$.

LEMMA 2.1.4: The function f is given by

$(2.1.29) \quad f(x,y) = x \log(x/y) + (1-x) \log\{(1-x)/(1-y)\}$ for all $(x,y) \in J$

Proof: Postulate $A5$ and $(2.1.16)$ give

$(2.1.30) \quad g(x) = -a [x \log(x/2) + (1-x) \log\{(1-x)/2\}] - bx \log\{x/(1-x)\}$,

for $x \in]0,1[$. Therefore $(2.1.16)$ and $(2.1.30)$ yield

$(2.1.31) \quad f(x,y) = a\{x \log x + (1-x) \log (1-x) - y \log y - (1-y) \log (1-y)\}$
$\qquad + b(x-y) \log\{y/(1-y)\}$, for $x,y \in]0,1[$.

Putting $x=2/3$, $y=1/3$ in $(2.1.31)$ and utilizing the postulate $A4$ it is seen that

$(2.1.32) \quad b = -1$

Again substituting $p_1=q_1=1/4$, $p_2=2/3$, $q_2=1/3$ in the second and third equation pair of $(2.1.14)$ one obtains

$(2.1.33) \quad f(1/4,1/4) + (3/4) f(8/9,4/9) = f(2/3,1/3) + (1/3) f(3/4,3/8)$

Thus $(2.1.31)$, $(2.1.32)$, $(2.1.33)$ and the postulates $A4$ and $A5$ yield

$(2.1.34) \quad a = 1$

Hence from $(2.1.31)$, $(2.1.32)$ and $(2.1.34)$ we have

$(2.1.35) \quad f(x,y) = x \log(x/y) + (1-x) \log\{(1-x)/(1-y)\}$, for $x,y \in]0,1[$.

Next we have to find $f(0,y)$ and $f(1,y)$ where $y \in]0,1[$. Taking $p_1=0$ in the second and third equation pair of $(2.1.14)$ we have

$(2.1.36) \quad f(0,q_1) + f(p_2,q_2/(1-q_1)) = f(p_2,q_2) + (1-p_2) f(0,q_1/(1-q_2))$

for $p_2,q_2 \in]0,1[$, $q_1 \in [0,1[$, $q_1+q_2 \in]0,1]$. For $q_1=1/2$, $q_2=1/4$, $(2.1.36)$ gives

Directed Divergence Between Two Distributions

(2.1.37) $f(0,1/2) + f(p_2,1/2) = f(p_2,1/4) + (1-p_2) f(0,2/3)$

for $p_2 \in]0,1[$. Equation (2.1.37) for $p_2=1/2$ and postulate A5 give

(2.1.38) $f(0,1/2) = f(1/2,1/4) + (1/2) f(0,2/3)$

Hence (2.1.35),(2.1.37) and (2.1.38) imply

(2.1.39) $f(0,1/2) = 1$

Taking $q_2=1-2q_1$ in (2.1.36) and utilizing (2.1.35) and (2.1.39) one gets from (2.1.36)

(2.1.40) $f(0,q_1) = - \log(1-q_1)$, for $q_1 \in]0,\frac{1}{2}[$

Again (2.1.36) with $q_2=1/2$ and (2.1.40) give

(2.1.41) $f(0,2q_1) = - \log(1-2q_1)$, for $2q_1 \in]0,1[$

Hence

(2.1.42) $f(0,q) = - \log(1-q)$, for $q \in]0,1[$

Also lemma 2.1.1 and (2.1.42) give

(2.1.43) $f(1,q) = - \log q$, for $q \in]0,1[$

Taking $q_1=0$, $p_1=0$ in the last two equations of (2.1.14) we easily find that $f(0,0)=0$ and hence from (2.1.13) we get

(2.1.44) $f(0,0) = f(1,1) = 0$

Now (2.1.35),(2.1.42),(2.1.43) and (2.1.44) complete the proof.

THEOREM 2.1.1: The function $I_n(P:Q)$ satisfying postulates A1 to A5 is uniquely determined as $\Sigma\, p_i \log(p_i/q_i)$.

Proof: By the repeated application of postulate A1 we have

(2.1.45) $I_n(P:Q) = \Sigma_{i=2}^n r_i\, I_2(r_{i-1}/r_i,\, p_i/r_i:\, s_{i-1}/s_i,\, q_i/s_i)$

where $r_i = p_1+\ldots+ p_i$ and $s_i = q_1+\ldots+ q_i$. Hence (2.1.45) and (2.1.29) give

$I_n(P:Q)$

$= \Sigma_{i=2}^n r_i[(p_i/r_i)\, \log(p_i s_i/r_i q_i) + (r_{i-1}/r_i)\log(r_{i-1}s_i)/(r_i s_{i-1})]$

$= \Sigma_{i=2}^n p_i\, \log(p_i/q_i) + \Sigma_{i=2}^n [r_{i-1}\log(r_{i-1}/s_{i-1}) - r_i\, \log(r_i/s_i)]$

$= \Sigma_{i=2}^n p_i\, \log(p_i/q_i) + r_1\, \log(r_1/s_1) - r_n \log(r_n/s_n)$

$= \Sigma_{i=1}^n p_i\, \log(p_i/q_i)$

and this completes the proof. It may be noticed that in (2.1.1) we use the convention $0 \log 0 = 0$ but nowhere in the proof of the theorem this condition is used. From the proof of theorem 2.1.1 it is evident that the directed divergence function $f(x,y) = I_2(x, 1-x: y, 1-y)$ has played a very important role. Also equation (2.1.14) satisfied by f is the main equation used in the proof of theorem 2.1.1. Therefore it may be regarded as a fundamental equation which may also be called *directed divergence functional equation*. This was listed as property *(viii)*.

Next we give a characterization theorem for directed divergence (2.1.1) by using a maximization principle. Before stating the theorem we will prove another theorem which is useful here as well as in later sections. This theorem is taken from Rathie(1971).

THEOREM 2.1.2: For $P \in S_n^*$, $Q \in S_n^*$ and $n > 2$, every solution of the functional inequality

(2.1.46) $\quad \sum_{i=1}^{n} g(p_i) f(p_i) \geq \sum_{i=1}^{n} g(p_i) f(q_i)$

is differentiable everywhere in $]0,1[$, where g is positive, strictly monotonic increasing and continuous in $]0,1[$. The general solution is given by

(2.1.47) $\quad f(p) = a \int dp/g(p) + b$

where $a \geq 0$ and b are arbitrary constants. Conversely, (2.1.47) satisfies (2.1.46).

Proof: The proof of theorem 2.1.2 depends on the following three lemmas.

LEMMA 2.1.5: f is monotonic increasing in $]0,1[$.

Proof: Putting $p_3 = q_3, \ldots, p_n = q_n$ in (2.1.46), we get

(2.1.48) $\quad g(p_1) f(p_1) + g(p_2) f(p_2) \geq g(p_1) f(q_1) + g(p_2) f(q_2)$

for $p_1 + p_2 = q_1 + q_2 < 1$. Since g is positive, (2.1.48) may be rewritten as

(2.1.49) $\quad [g(p_1)/g(p_2)][f(p_1) - f(q_1)] \geq f(q_2) - f(p_2)$

Due to symmetry, interchanging p_1 and q_1 and also p_2 and q_2 in (2.1.49) we get

(2.1.50) $\quad [g(q_1)/g(q_2)][f(q_1) - f(p_1)] \geq f(p_2) - f(q_2)$

Adding (2.1.49) and (2.1.50) we have

(2.1.51) $\quad [\{g(p_1)/g(p_2)\} - \{g(q_1)/g(q_2)\}][f(p_1) - f(q_1)] \geq 0$

Let $p_1 > q_1$, then $p_2 < q_2$. Since g is strictly monotonic increasing $g(p_1) > g(q_1)$ and $g(p_2) < g(q_2)$, and thus

(2.1.52) $\qquad [g(p_1)/g(p_2)] > [g(q_1)/g(q_2)]$

Hence (2.1.51) and (2.1.52) yield $f(p_1) \geqslant f(q_1)$ for $p_1 > q_1$ proving the lemma 2.1.5.

LEMMA 2.1.6: For an arbitrary constant $a \geqslant 0$, $g(p)f'(p) = a$ holds good for all those points $p \in]0,1[$ at which f is differentiable, where f' denotes the derivative of f.

Proof: f is almost everywhere differentiable in $]0,1[$ because it is monotonic increasing in $]0,1[$ (see Natanson(1964)). Now rewriting (2.1.48) we get

(2.1.53) $\qquad g(p_1)[f(q_1)-f(p_1)] \leqslant g(p_2)[f(p_2)-f(q_2)]$

Substituting $q_1 = p_1 + \delta$, $q_2 = p_2 - \delta$, $\delta > 0$ in (2.1.53) and then dividing by δ, we have

(2.1.54) $\qquad g(p_1)[f(p_1+\delta)-f(p_1)]/\delta \leqslant g(p_2)[f(p_2)-f(p_2-\delta)]/\delta$

Let p_1 and p_2 be two points in $]0,1[$ where f is differentiable. Then taking $\delta \to 0$ in (2.1.54) one gets

(2.1.55) $\qquad g(p_1)f'(p_1) \leqslant g(p_2)f'(p_2)$

Interchanging the roles of p_1 and p_2 in (2.1.55) we get

(2.1.56) $\qquad g(p_2)f'(p_2) \leqslant g(p_1)f'(p_1)$

Combining (2.1.55) and (2.1.56), we have

(2.1.57) $\qquad g(p_1)f'(p_1) = g(p_2)f'(p_2)$

Hence we have $g(p)f'(p) = a$, where a is a constant. Since f is monotonic increasing $f'(p) \geqslant 0$. Also $g > 0$ and hence $a \geqslant 0$.

LEMMA 2.1.7: The function f is differentiable everywhere in $]0,1[$.

Proof: Let p be an arbitrary point in $]0,1[$ and let f'_-, f'_+, f'^-, f'^+ denote the four Dini-derivatives at p. Let f be differentiable at a point p_1. Then (2.1.54) with $p_2 = p$ and taking $\inf \delta \to 0$ gives

(2.1.58) $\qquad a \leqslant g(p)f'_-(p)$

Also taking $p_1 = p$ in (2.1.54) and assuming f to be differentiable at p_2 and on taking $\sup \delta \to 0$, we have

(2.1.59) $g(p)f'^{+}(p) \leq a$

Again taking $p_1 = q_1+\delta$, $p_2 = q_2-\delta$, $\delta > 0$ in (2.1.53) and dividing by δ we get

(2.1.60) $f(p_1+\delta)[f(q_1+\delta)-f(q_1)]/\delta = g(q_2-\delta)[f(q_2)-f(q_2-\delta)]/\delta$

So if f is differentiable at q_1 and if $q_2 = p$, then taking $sup\ \delta \to 0$ and assuming that g is continuous, we get

(2.1.61) $a \geq g(p)f'^{-}(p)$

Finally taking $q_1 = p$, assuming f is differentiable at q_2 in (2.1.60) and taking $inf\ \delta \to 0$, we have

(2.1.62) $g(p)f'_{+}(p) \geq a$

Hence (2.1.58),(2.1.59),(2.1.61) and (2.1.62) give

(2.1.63) $f'_{-}(p) = f'^{-}(p) = f'_{+}(p) = f'^{+}(p) = a/g(p)$

showing that f is differentiable everywhere in $]0,1[$ and that $g(p)f'(p) = a$. Thus the "if" part of theorem 2.1.2 is proved. For the "only if" part we have to show that

(2.1.64) $\sum_{p_i>q_i} g(p_i) \int_{q_i}^{p_i} dx/g(x) \geq \sum_{q_i>p_i} g(p_i) \int_{p_i}^{q_i} dx/g(x)$

Since $g(x)$ is monotonic increasing $1/g(x)$ is monotonic decreasing and hence

$$L.H.S. \text{ of } (2.1.64) \geq \sum_{p_i>q_i} g(p_i)(p_i-q_i)/g(p_i) = \sum_{p_i>q_i}(p_i-q_i),$$

and

$$R.H.S. \text{ of } (2.1.64) \leq \sum_{q_i>p_i} g(p_i)(q_i-p_i)/g(p_i) = \sum_{q_i>p_i}(q_i-p_i).$$

Since $\sum_{p_i>q_i}(p_i-q_i) = \sum_{q_i>p_i}(q_i-p_i)$, (2.1.64) holds. This completes the proof of theorem 2.1.2.

Now two important special cases of theorem 2.1.2 will be discussed, due to their usefulness for characterization purposes. When $g(p)=p$ theorem 2.1.2 takes the following form:

Corollary 2.1.1: The general solution of the functional inequality

(2.1.65) $\sum_{i=1}^{n} p_i f(p_i) \geq \sum_{i=1}^{n} p_i f(q_i)$

for $P \in S_n^*$ and $Q \in S_n^*$, $n > 2$, is given by

(2.1.66) $f(x) = a \log x + b$

where $a \geq 0$ and b are arbitrary real constants.

Aczél and Pfanzagl (1966) obtained all the differentiable solutions of (2.1.65). Muszély (1973) obtained the continuous solutions of the functional inequality (2.1.65) for $n=2$ while Fischer (1972) gave the general solutions without any assumption of regularity on f. When $g(p) = p^\beta$, theorem 2.1.2 reduces to the following form.

Corollary 2.1.2: Every solution of the functional inequality

(2.1.67) $\sum p_i^\beta f(p_i) \geq \sum p_i^\beta f(q_i)$

for $P \in S_n^*$, $Q \in S_n^*$ and $n > 2$ is given by

(2.1.68) $f(x) = a(\beta) x^{1-\beta}/(1-\beta) + b$, $\beta > 0$, $\beta \neq 1$

where $a(\beta) \geq 0$ and b are arbitrary real constants.

The result contained in corollary 2.1.2 as well as other related results were proved earlier by Rathie (1973) and it will be used later in this chapter. Next we prove another characterization theorem for directed divergence (2.1.1) by using a maximization principle.

THEOREM 2.1.3: The function $I_n(P:Q)$ for $P \in S_n^*$, $Q \in S_n^*$, and for a fixed $n > 2$, satisfying the postulates:

B1. $I_n(P:Q) = \sum_{i=1}^{n} p_i [f(p_i) - f(q_i)]$

B2. $I_n(P:Q) \geq 0$

B3. $I_n(1/8, 1/2, 3/8(n-2), \ldots, 3/8(n-2) : 1/2, 1/8, 3/8(n-2), \ldots, 3/8(n-2)) = 3/4$

is the directed divergence (2.1.1).

Proof: Postulates B1 and B2 are equivalent to (2.1.65) of corollary 2.1.1 and hence from (2.1.66) and B1 we have

(2.1.69) $I_n(P:Q) = a \sum p_i \log(p_i/q_i)$

Now (2.1.69) and postulate B3 give $a=1$, thus proving theorem 2.1.3.

2.2 DIRECTED DIVERGENCE OF ORDER α

For the two distributions $P = (p_1,\ldots,p_n) \in S_n$ and $Q = (q_1,\ldots,q_n) \in S_n$ the following two generalized measures of directed divergence involving a parameter are defined.

(a) DEFINITION:

Various quantities which can be considered to be generalizations of directed divergence defined in section 2.1 will be defined here.

Definition 2.2.1: Additive directed divergence of order α

(2.2.1) $\hat{I}_{n,\alpha}(P:Q) = (\alpha-1)^{-1} \log(\Sigma\, p_i^\alpha q_i^{1-\alpha})$, $\alpha \neq 1$

Definition 2.2.2: Non-additive directed divergence of order α

(2.2.2) $I_{n,\alpha}(P:Q) = (\Sigma\, p_i^\alpha q_i^{1-\alpha} -1)/(2^{\alpha-1}-1)$, $\alpha \neq 1$

In (2.2.1) and (2.2.2) we use the convention $0^\alpha = 0$, $(\alpha \neq 0)$. Clearly from (2.2.1) and (2.2.2) we get the following relation between the two types of directed divergences.

(2.2.3) $I_{n,\alpha} = (2^{(\alpha-1)\hat{I}_{n,\alpha}} -1)/(2^{\alpha-1}-1)$, $\alpha \neq 1$

For $n=2$, (2.2.2) reduces to the following form,

(2.2.4) $I_{2,\alpha}(p,1-p:q,1-q) = [p^\alpha q^{1-\alpha} + (1-p)^\alpha (1-q)^{1-\alpha} -1]/(2^{\alpha-1}-1)$, $\alpha \neq 1$

and $p,q \in [0,1]$.

As will be seen later in this section, the quantity in (2.2.4) plays a key role in the characterization theorems. Hence we will call $I_{2,\alpha}(p,1-p:q,1-q)$ as *non-additive directed divergence function of order* α. We may also define (2.2.2) and (2.2.4) alternately as follows.

Definition 2.2.3: A function $g: I \times I \to \mathcal{R}$ where $I = [0,1]$ and $\mathcal{R}$ the real numbers, is called a non-additive directed divergence function of order α ($\alpha \neq 1$) provided g satisfies the functional equation

(2.2.5) $g(x,y) + (1-x)^\alpha (1-y)^{1-\alpha} g(u/(1-x), v/(1-y))$
$= g(u,v) + (1-u)^\alpha (1-v)^{1-\alpha} g(x/(1-u), y/(1-v))$

for all $x,y,u,v \in [0,1[$ with $x+u, y+v \in [0,1]$ and with the boundary conditions,

(2.2.6) $g(0,0) = g(1,1)$

Directed Divergence of Order α

and

(2.2.7) $g(0,\frac{1}{2}) = g(1,\frac{1}{2}) = 1$

Later on we will show that $g(p,q)$ is uniquely determined as $I_{2,\alpha}(p,1-p:1-q)$ given by (2.2.4).

Definition 2.2.4: If g is a directed divergence function of order α ($\alpha \neq 1$) as defined above, then we define the non-additive directed divergence of order α by the relation

(2.2.8) $\hat{I}_{n,\alpha}(P:Q) = \sum_{i=2}^{n} r_i^{\alpha} s_i^{1-\alpha} g(p_i/r_i, q_i/s_i)$,

where $r_i = p_1 + \ldots + p_i$, $s_i = q_1 + \ldots + q_i$ for $i=1,2,\ldots,n$. It is interesting to notice that as $\alpha \to 1$ we have

(2.2.9) $\lim_{\alpha \to 1} \hat{I}_{n,\alpha} = I_n$, $\lim_{\alpha \to 1} I_{n,\alpha} = I_n$

for fixed n, where I_n is given in (2.1.1). In this sense they may be regarded as generalizations of directed divergence defined in section 2.1.

(b) PROPERTIES:

As in section 2.1 some of the properties of generalized measure of directed divergence are listed below.

(i)   Non-negativity: $\hat{I}_{n,\alpha}(P:Q)$ and $I_{n,\alpha}(P:Q)$ are both non-negative and they are zeros if and only if $p_i = q_i$ for all i.

(ii)   Symmetry:

(2.2.10) $\hat{I}_{n,\alpha}(p_1,\ldots,p_n:q_1,\ldots,q_n) = \hat{I}_{n,\alpha}(p_{a(1)},\ldots,p_{a(n)}:q_{a(1)},\ldots,q_{a(n)})$

where $\{a(1),\ldots,a(n)\}$ is an arbitrary permutation of $\{1,2,\ldots,n\}$. The same property holds good for $I_{n,\alpha}$.

(iii) Expansibility:

(2.2.11) $\hat{I}_{n+1,\alpha}(P,0:Q,0) = \hat{I}_{n,\alpha}(P:Q)$

Same property holds good for $I_{n,\alpha}$.

(iv)   Recursivity or Branching Principle:

(2.2.12) $I_{n,\alpha}(p_1,\ldots,p_n:q_1,\ldots,q_n) = I_{n-1,\alpha}(p_1+p_2,p_3,\ldots,p_n:q_1+q_2,q_3,\ldots,q_n)$

$\quad + (p_1+p_2)^{\alpha}(q_1+q_2)^{1-\alpha} I_{2,\alpha}(p_1/(p_1+p_2),p_2/(p_1+p_2):q_1/(q_1+q_2),$

$\quad q_2/(q_1+q_2))$, for p_1+p_2, $q_1+q_2 > 0$.

(v)   Additivity:

(2.2.13) $\hat{I}_{mn,\alpha}(p_1P_1,\ldots,p_1P_m,\ldots,p_nP_1,\ldots,p_nP_m:q_1Q_1,\ldots,q_1Q_m,\ldots,q_nQ_1,\ldots,q_nQ_m)$

$= \hat{I}_{n,\alpha}(p_1,\ldots,p_n:q_1,\ldots,q_n) + \hat{I}_{m,\alpha}(P_1,\ldots,P_m:Q_1,\ldots,Q_m)$

where $(p_1,\ldots,p_n) \in S_n$, $(q_1,\ldots,q_n) \in S_n$, $(P_1,\ldots,P_m) \in S_m$ and $(Q_1,\ldots,Q_m) \in S_m$.

(vi)   Non-additivity:

For $P \in S_n$, $Q \in S_n$, $(P_1,\ldots,P_m) \in S_m$ and $(Q_1,\ldots,Q_m) \in S_m$,

(2.2.14) $I_{mn,\alpha}(p_1P_1,\ldots,p_1P_m,\ldots,p_nP_1,\ldots,p_nP_m:q_1Q_1,\ldots,q_1Q_m,\ldots,q_nQ_1,\ldots,q_nQ_m)$

$= I_{n,\alpha}(p_1,\ldots,p_n:q_1,\ldots,q_n) + I_{m,\alpha}(P_1,\ldots,P_m:Q_1,\ldots,Q_m)$

$+ (2^{\alpha-1}-1)I_{n,\alpha}(p_1,\ldots,p_n:q_1,\ldots,q_n)I_{m,\alpha}(P_1,\ldots,P_m:Q_1,\ldots,Q_m)$

(vii)   Strong Non-additivity:

For $P \in S_n$, $Q \in S_n$, $(P_{1i},\ldots,P_{mi}) \in S_m$, $(Q_{1i},\ldots,Q_{mi}) \in S_m$, $i=1,\ldots,n$,

(2.2.15) $I_{mn,\alpha}(p_1P_{11},p_1P_{21},\ldots,p_1P_{m1},\ldots,p_nP_{1n},p_nP_{2n},\ldots,p_nP_{mn}:q_1Q_{11},q_1Q_{21}$

$\ldots,q_1Q_{m1},\ldots,q_nQ_{1n},q_nQ_{2n},\ldots,q_nQ_{mn})$

$= I_{n,\alpha}(p_1,\ldots,p_n:q_1,\ldots,q_n) +$

$+ \sum_{i=1}^{n} p_i^\alpha q_i^{1-\alpha} I_{m,\alpha}(P_{1i},\ldots,P_{mi}:Q_{1i},\ldots,Q_{mi})$

(viii)   Continuity: Both $\hat{I}_{n,\alpha}$ and $I_{n,\alpha}$ are continuous functions of their *2n* variables.

(c)   CHARACTERIZATION THEOREMS:

Here we will state and prove four theorems concerning non-additive measures of directed divergence of order α given in definition *2.2.2* and its special case for *n=2*. Corresponding results for additive directed divergence given in definition *2.2.1* can be obtained by using the relation *(2.2.3)* and hence they are not discussed separately. The following theorem is due to Rathie and Kannappan (1972).

THEOREM 2.2.1: If *g* is a non-additive directed divergence function of order α for $\alpha \neq 1$, that is, if *g* is a solution of the functional equation *(2.2.5)* satisfying *(2.2.6)* and *(2.2.7)*, then *g(x,y)* has the form *(2.2.4)* for all $x,y \in [0,1]$.

Directed Divergence of Order α   49

Proof: For $x = 0$, $y = 0$, *(2.2.5)* gives

(2.2.16) $[1-(1-u)^{\alpha}(1-v)^{1-\alpha}]g(0,0) = 0$

for all $u,v \in [0,1[$, which in turn implies $g(0,0)=0$. Therefore, from *(2.2.6)* we have

(2.2.17) $g(0,0) = g(1,1) = 0$

Taking $u = 1-x$, $v = 1-y$ in *(2.2.5)* and then applying *(2.2.17)* we obtain

(2.2.18) $g(x,y) = g(1-x,1-y)$, for all $x,y \in]0,1[$.

Taking $p = u/(1-x)$, $q = v/(1-y)$, $r = 1-x$ and $s = 1-y$ in *(2.2.5)*, we have

(2.2.19) $g(1-r,1-s) + r^{\alpha}s^{1-\alpha} g(p,q) = g(pr,qs)$
$$+ (1-pr)^{\alpha}(1-qs)^{1-\alpha} g((1-r)/(1-pr),(1-s)/(1-qs))$$

for all $p,q \in [0,1]$, $r,s \in]0,1]$ such that $pr \neq 1$, $qs \neq 1$. The equation *(2.2.19)* with $r=1, s=\frac{1}{2}$ gives

(2.2.20) $g(0,\frac{1}{2}) + 2^{\alpha-1} g(p,q) = g(p,q/2) + (1-p)^{\alpha}(1-q/2)^{1-\alpha} g(0,1/(2-q))$

for $p \in [0,1[$, $q \in [0,1]$. Taking $q=0$ in *(2.2.20)* and using *(2.2.7)* we have

(2.2.21) $g(p,0) = [(1-p)^{\alpha} -1](2^{\alpha-1} -1)^{-1}$,

for $p \in [0,1[$. The equation *(2.2.19)* for $p=1$, $q=0$ yields

(2.2.22) $g(1-r,1-s) + r^{\alpha}s^{1-\alpha} g(1,0) = g(r,0) + (1-r)^{\alpha} g(1,1-s)$

for $r \in]0,1[$ $s \in]0,1]$. Now *(2.2.21)* and *(2.2.22)* for $s=1$ give

(2.2.23) $g(1,0) = (1-2^{\alpha-1})^{-1}$

Again putting $r=1$, $q=0$ in *(2.2.19)*, one gets

(2.2.24) $g(0,1-s) + s^{1-\alpha} g(p,0) = g(p,0) + (1-p)^{\alpha} g(0,1-s)$

which on using *(2.2.21)* gives

(2.2.25) $g(0,1-s) = (s^{1-\alpha} -1)(2^{\alpha-1} -1)^{-1}$

$s \in]0,1]$. The equation *(2.2.19)* for $p=0$, $q=1$ takes the following form:

(2.2.26) $g(1-r,1-s) + r^{\alpha}s^{1-\alpha} g(0,1) = g(0,s) + (1-s)^{1-\alpha} g(1-r,1)$

for $r\in]0,1]$, $s\in]0,1[$. Now (2.2.26) with $r=1$ and (2.2.25) give

(2.2.27) $g(0,1) = (1-2^{\alpha-1})^{-1}$

From (2.2.22) and (2.2.26) on using (2.2.23) and (2.2.27) we have

(2.2.28) $g(r,0) + (1-r)^{\alpha} g(1,1-s) = g(0,s) + (1-s)^{1-\alpha} g(1-r,s)$

for $r,s\in]0,1[$. The equation (2.2.28) for $s=1/2$, (2.2.21) and (2.2.7) give

(2.2.29) $g(1-r,1) = [(1-r)^{\alpha} -1]/(2^{\alpha-1} -1)$

for $r\in]0,1[$. Hence from (2.2.21), (2.2.25), (2.2.28) and (2.2.29) we have

(2.2.30) $g(1,1-s) = [(1-s)^{1-\alpha} -1]/(2^{\alpha-1} -1)$

for $s\in]0,1[$. Also (2.2.26), (2.2.27), (2.2.29) and (2.2.25) give

(2.2.31) $g(1-r,1-s) = [r^{\alpha}s^{1-\alpha} + (1-r)^{\alpha}(1-s)^{1-\alpha} -1]/(2^{\alpha-1} -1)$

for $r,s\in]0,1[$. Now we see from (2.2.21), (2.2.23), (2.2.25), (2.2.27), (2.2.29), (2.2.30) and (2.2.31) that g is given by (2.2.4) and that (2.2.18) is true for all $x,y\in[0,1]$. This completes the proof.

THEOREM 2.2.2: Let g be a non-additive directed divergence function of order $\alpha(\alpha\neq 1)$ and $I_{n,\alpha}$ be the non-additive directed divergence of order α. Then $I_{n,\alpha}$ is given by (2.2.2).

Proof: By theorem 2.2.1, g has the form (2.2.4). Now (2.2.8) becomes

$$I_{n,\alpha}(P:Q) = (2^{\alpha-1} -1)^{-1} \cdot \Sigma_{i=2}^{n} r_i^{\alpha} s_i^{1-\alpha}[(p_i/r_i)^{\alpha}(q_i/s_i)^{1-\alpha}$$
$$+ (1-p_i/r_i)^{\alpha}(1-q_i/s_i)^{1-\alpha} -1]$$
$$= (2^{\alpha-1} -1)^{-1} [\Sigma_{i=2}^{n} p_i^{\alpha} q_i^{1-\alpha}) + r_1^{\alpha}s_1^{1-\alpha} - r_n^{\alpha}s_n^{1-\alpha}]$$
$$= (2^{\alpha-1} -1)^{-1} (\Sigma_{i=1}^{n} p_i^{\alpha} q_i^{1-\alpha} -1)$$

which is precisely (2.2.2).

THEOREM 2.2.3: If $I_{n,\alpha}$ for $n=2,3,\ldots$ satisfies the following postulates:

C1. $I_{n,\alpha}(p_1,\ldots,p_n:q_1,\ldots,q_n) = I_{n-1,\alpha}(p_1+p_2,p_3,\ldots,p_n:q_1+q_2,q_3,\ldots,q_n)$

$$+ (p_1+p_2)^\alpha (q_1+q_2)^{1-\alpha} I_{2,\alpha}(p_1/(p_1+p_2),p_2/(p_1+p_2):q_1/(q_1+q_2),$$
$$q_2/(q_1+q_2)), \text{ with } p_1+p_2, \; q_1+q_2 > 0$$

C2. $I_{3,\alpha}(p_1,p_2,p_3:q_1,q_2,q_3)$ is a symmetric function of its variables $\{p_i,q_i\}$, $i = 1,2,3$

C3. $I_{2,\alpha}(1,0:\tfrac{1}{2},\tfrac{1}{2}) = 1$

then $I_{n,\alpha}$ is uniquely determined as given by (2.2.2).

Proof: We define

(2.2.32) $\quad g(x,y) = I_{2,\alpha}(x,1-x:y,1-y)$, $x,y \in [0,1]$

Then we will show that g is a non-additive directed divergence function of order α. For $n=3$, the postulate C1 takes the following form:

(2.2.33) $\quad I_{3,\alpha}(p_1,p_2,p_3:q_1,q_2,q_3) = I_{2,\alpha}(p_1+p_2,p_3:q_1+q_2,q_3)$
$$+ (p_1+p_2)^\alpha (q_1+q_2)^{1-\alpha} I_{2,\alpha}(p_1/(p_1+p_2),p_2/(p_1+p_2):q_1/(q_1+q_2),q_2/(q_1+q_2))$$

for p_1+p_2, $q_1+q_2 > 0$. Interchanging p_1 and p_2 and q_1 and q_2 in (2.2.33) and using the postulate C2, (2.2.33) gives

(2.2.34) $\quad I_{2,\alpha}(p_1/(p_1+p_2),p_2/(p_1+p_2):q_1/(q_1+q_2),q_2/(q_1+q_2))$
$$= I_{2,\alpha}(p_2/(p_1+p_2),p_1/(p_1+p_2):q_2/(q_1+q_2),q_1/(q_1+q_2))$$

for $p_1,p_2,q_1,q_2 \in [0,1]$ with p_1+p_2, $q_1+q_2 \in \,]0,1]$. Hence (2.2.32), (2.2.34) give

(2.2.35) $\quad g(x,y) = g(1-x,1-y)$, for all $x,y \in [0,1]$.

From the postulate C2, we get

(2.2.36) $\quad I_{3,\alpha}(x,1-x-u,u:y,1-y-v,v) = I_{3,\alpha}(u,1-x-u,x:v,1-y-v,y)$

where $x,y,u,v,1-x-u,1-y-v \in [0,1]$. Now (2.2.36) on utilizing the postulate C1 for $n=3$ and (2.2.32) gives

(2.2.37) $\quad g(1-u,1-v) + (1-u)^\alpha (1-v)^{1-\alpha} g(x/(1-u),y/(1-v))$
$$= g(1-x,1-y) + (1-x)^\alpha (1-y)^{1-\alpha} g(u/(1-x),v/(1-y))$$

for $x,y,u,v \in [0,1[$ with $x+u$, $y+v \in [0,1]$. Due to (2.2.35), (2.2.37) takes the form

$(2.2.38)$ $g(u,v) + (1-u)^{\alpha}(1-v)^{1-\alpha} g(x/(1-u),y/(1-v))$

$$= g(x,y) + (1-x)^{\alpha}(1-y)^{1-\alpha} g(u/(1-x),v/(1-y))$$

which is precisely $(2.2.5)$. Also the postulate $C2$ on using the postulate $C1$ and $(2.2.32)$ gives $I_{3,\alpha}(1,0,0:1,0,0) = 2 g(1,1)$ and $I_{3,\alpha}(0,1,0:0,1,0) = g(1,1) + g(0,0)$. Hence

$(2.2.39)$ $g(1,1) = g(0,0)$

Substituting $x=y=1/2$, $u=v=0$ in $(2.2.38)$ we get

$(2.2.40)$ $g(1,1) = (1/2) g(0,0)$

Thus from $(2.2.39)$ and $(2.2.40)$ we have

$(2.2.41)$ $g(1,1) = g(0,0) = 0$

which is $(2.2.17)$. Further, from the postulate $C3$, $(2.2.32)$ and $(2.2.35)$ for $x=0$ and $y=1/2$, one gets

$(2.2.42)$ $g(1,\tfrac{1}{2}) = g(0,\tfrac{1}{2}) = 1$

which is $(2.2.7)$. Hence from $(2.2.38)$, $(2.2.41)$ and $(2.2.42)$ we conclude that g is a non-additive directed divergence function of order α. So by theorem $2.2.1$, g is given by $(2.2.4)$. Now theorem $2.2.3$ easily follows by induction from the postulate $C1$, $(2.2.32)$ and $(2.2.4)$.

Remark: By using $(2.2.3)$ these theorems can be restated for additive directed divergence of order α, that is, for $(2.2.1)$ as well.

THEOREM 2.2.4: The function $\hat{I}_{n,\alpha}(P:Q)$ for $P \varepsilon S_n$ and $Q \varepsilon S_n$ and for a fixed $n>2$ satisfying the postulates:

D1. $\hat{I}_{n,\alpha}(P:Q) = \Sigma [p_i^{\alpha} g(p_i) - g(q_i)]$, $\alpha > 0$, $\alpha \neq 1$

D2. $\hat{I}_{n,\alpha}(P:Q) \geqslant 0$ is uniquely determined as

$\hat{I}_{n,\alpha}(P:Q) = \dfrac{a(\alpha)}{1-\alpha} [\Sigma p_i^{\alpha} q_i^{1-\alpha} -1]$, $1 > \alpha > 0$, $a(\alpha) \geqslant 0$

The proof of theorem $2.2.4$ follows from corollary $2.1.2$. The constant $a(\alpha)$ can be easily determined if we take one normalization postulate. In $(2.2.2)$ we use the convention $0^{\alpha}=0$ $(\alpha \neq 0)$ for convenience but nowhere in the proof this condition is used.

Directed Divergence of Order α

The following theorem is due to Kannappan & Rathie (1973c)

THEOREM 2.2.5: The function $F:[0,1]\times[0,1]\to \mathcal{R}$, measurable in each variable and satisfying the functional equation

(2.2.43) $\quad F(x,y) + (1-x)^{\alpha}(1-y)^{\gamma} F(u/(1-x), v/(1-y))$

$\qquad = F(u,v) + (1-u)^{\alpha}(1-v)^{\gamma} F(x/(1-u), y/(1-v))$

for $x,y,u,v \in [0,1[$ with $x+u, y+v \in [0,1]$ is given by

(2.2.44) $\quad F(x,y) = b\, x^{\alpha} y^{\gamma} - A[x^{\alpha} y^{\gamma} + (1-x)^{\alpha}(1-y)^{\gamma} - 1]$

for $x,y \in [0,1]$, $\alpha(>0) \neq 1$, where A and b are arbitrary constants.

Proof: For each fixed $y,v \in [0,1[$ with $y+v \in [0,1]$, the functional equation (2.2.43) is of the form (1.3.11). Hence applying the results of theorem 1.3.2, we find that there exist constants $A(y,v)$, $d_1(y,v)$, $d_2(y,v)$, $c_1(y,v)$, $c_2(y,v)$ and $c_4(y,v)$ such that

(2.2.45) $\quad F(x,y) = A(y,v) f_{\alpha}(x) + d_1(y,v) x^{\alpha} + c_1(y,v) - c_2(y,v)(1-x)^{\alpha}$

(2.2.46) $\quad (1-y)^{\gamma} F(x,v/(1-y)) = A(y,v) f_{\alpha}(x) + d_2(y,v) x^{\alpha} + c_2(y,v)$

(2.2.47) $\quad F(x,v) = A(y,v) f_{\alpha}(x) + d_2(y,v) x^{\alpha} + c_1(y,v) - c_4(y,v)(1-x)^{\alpha}$

(2.2.48) $\quad (1-v)^{\gamma} F(x,y/(1-v)) = A(y,v) f_{\alpha}(x) + d_1(y,v) x^{\alpha} + c_4(y,v)$

where $f_{\alpha}(x)$ is given by (1.2.5).

From (2.2.45) it is easy to find that $A(y,v) + d_1(y,v)$, $A(y,v) - c_2(y,v)$ and $A(y,v) - c_1(y,v)$ are all functions of y alone. Also from (2.2.47) we see that $A(y,v) - c_1(y,v)$ is a function of v alone. Hence $-A(y,v) + c_1(y,v)$ is a constant, say A. Denoting $A(y,v)+d_1(y,v)$ and $A(y,v) - c_2(y,v)$ by $B(y)$ and $C(y)$ respectively, the equation (2.2.45) becomes,

(2.2.49) $\quad F(x,y) = A + B(y) x^{\alpha} + C(y)(1-x)^{\alpha}$, for $x,y \in [0,1]$.

Now, substituting the expression for F from (2.2.49) in (2.2.43) we have

(2.2.50) $\quad C(y) = -A (1-y)^{\gamma} \quad$ and

(2.2.51) $\quad B(y) - (1-v)^{\gamma} B(y/(1-v)) = 0$

The equation (2.2.51) yields

(2.2.52) $\quad B(y) = \lambda y^\gamma$

where λ is an arbitrary constant. Finally, taking $b=\lambda+A$ and substituting for $B(y)$ and $C(y)$ from (2.2.52) and (2.2.50) in (2.2.49) we arrive at (2.2.44). Thus the proof of the theorem 2.2.5 is complete.

Corollary 2.2.1: For $\gamma = 1-\alpha$, $F(1,1) = F(0,0)$, $F(0,\tfrac{1}{2}) = 1$ the function F of (2.2.44) reduces to the directed divergence function of type α given by (2.2.4).

2.3 SOME GENERALIZATIONS

In this section we list some of the generalizations of the various quantities discussed in the last two sections.

(a) DEFINITIONS

(i) Generalization due to Kapur (1968)

(2.3.1) $\quad I_{n,1}^\beta(P:Q) = \Sigma\, p_i^\beta \log(p_i/q_i)/(\Sigma\, p_i^\beta)$

(2.3.2) $\quad I_{n,\alpha}^\beta(P:Q) = (\alpha-1)^{-1} \log(\Sigma\, p_i^{\alpha+\beta-1} q_i^{1-\alpha}/\Sigma\, p_i^\beta)$, $\alpha \neq 1$

(ii) Generalizations due to Rathie (1971a, 1971b, 1971c)

(2.3.3) $\quad I_{n,1}(p_1,\ldots,p_n;\beta_1,\ldots,\beta_n;q_1,\ldots,q_n) = \Sigma\, p_i^{\beta_i} \log(p_i/q_i)/(\Sigma\, p_i^{\beta_i})$

(2.3.4) $\quad I_{n,\alpha}(p_1,\ldots,p_n;\beta_1,\ldots,\beta_n;q_1,\ldots,q_n)$

$\qquad = (\alpha-1)^{-1} \log(\Sigma\, p_i^{\alpha+\beta_i-1} q_i^{1-\alpha}/\Sigma\, p_i^{\beta_i})$, $\alpha \neq 1$

In addition to the above quantities, the following are also discussed.

(2.3.5) $\quad \underline{I}_{n,\alpha}(p_1,\ldots,p_n;\beta_1,\ldots,\beta_n;q_1,\ldots,q_n)$

$\qquad = (\Sigma\, p_i^{\alpha+\beta_i-1} q_i^{1-\alpha}/p_i^{\beta_i} -1)/(2^{\alpha-1}-1)$, $\alpha \neq 1$

and its special case when $\beta_i = \beta$ for all i.

(2.3.6) $\quad I_n^\beta <p_1,\ldots,p_n;q_1,\ldots,q_n> = \Sigma\, p_i^{\beta+1} \log(p_i/q_i)$

A number of properties and several results involving the above quantities may be found in the list of references given at the end of this book. As some of the properties are similar to what we have discussed in the previous sections further discussion is omitted. Some characterization theorems will be mentioned in

Directed Divergences Involving More Than Two Distributions

chapter 4, under the conditions $\Sigma p_i \leq a$ and $\Sigma q_i \leq a$ where $a=1$ or any other positive number.

2.4 DIRECTED DIVERGENCES INVOLVING MORE THAN TWO DISTRIBUTIONS

Let $P = (p_1,\ldots,p_n)$, $Q = (q_1,\ldots,q_n)$, $R = (r_1,\ldots,r_n)$, $p_i, q_i, r_i \geq 0$, $\Sigma p_i = 1 = \Sigma q_i = \Sigma r_i$ be three finite discrete probability distributions. That is, $P, Q, R \in S_n$. Then for these three distributions we define the following quantities.

(a) Definition 2.4.1: A generalized directed divergence is defined by the expression,

(2.4.1) $I_n(P:Q:R) = \Sigma\, p_i\, log(q_i/r_i)$

Here the convention $0\, log\, 0 = 0$ is followed. Also whenever q_i or r_i is zero then the corresponding p_i is also zero and $log(q_i/r_i)$ is to be taken as $(log\, q_i - log\, r_i)$.

Definition 2.4.2: A non-additive generalized directed divergence of order α is defined as follows:

(2.4.2) $I_{n,\alpha}(P:Q:R) = \Sigma(p_i q_i^{\alpha-1} r_i^{1-\alpha} -1)/(2^{\alpha-1} -1)$, $\alpha \neq 1$

We observe that (2.4.2) is a generalization of (2.4.1) since $\lim_{\alpha \to 1} I_{n,\alpha} = I_n$. Also (2.4.1) and (2.4.2) are generalizations of (2.1.1) and (2.2.2) respectively to which they reduce when $P=Q$. Some interesting special cases of (2.4.1) and (2.4.2) for $n=2$ are given below.

(2.4.3) $I_2(p, 1-p:q, 1-q:r, 1-r) = p\, log(q/r) + (1-p)\, log\{(1-q)/(1-r)\}$

for $(p,q,r) \in\,]0,1[\, \times]0,1[\, \times\,]0,1[\cup \{(0,y,z)\} \cup \{(1,y',z')\}$, $y, z \in [0,1[$, $y', z' \in\,]0,1]$

(2.4.4) $I_{2,\alpha}(p, 1-p:q, 1-q:r, 1-r)$

$= [pq^{\alpha-1} r^{1-\alpha} + (1-p)(1-q)^{\alpha-1}(1-r)^{1-\alpha} -1]/(2^{\alpha-1} -1)$

for $p,q,r \in [0,1]$. We will call (2.4.3) and (2.4.4) as generalized directed divergence function and non-additive generalized directed divergence function of order α respectively. Further, an additive generalized directed divergence of order α may be defined as follows.

Definition 2.4.3:

$(2.4.5) \quad \hat{I}_{n,\alpha}(P:Q:R) = (\alpha-1)^{-1} \log(\Sigma\, p_i q_i^{\alpha-1} r_i^{1-\alpha}),\ \alpha \neq 1$

We use the convention $0^\alpha = 0, (\alpha \neq 0)$ in (2.4.2), (2.4.4) and (2.4.5) but nowhere in the proof of the theorems $0^\alpha = 0$ will be used. Clearly (2.4.2) and (2.4.5) are connected by the relation,

$(2.4.6) \quad I_{n,\alpha} = [2^{(\alpha-1)\hat{I}_{n,\alpha}} - 1]/(2^{\alpha-1} - 1),\ \alpha \neq 1$

Alternate definitions of some of these quantities can be given through functional equations.

Definition 2.4.4: A real valued function h defined on $[0,1] \times [0,1] \times [0,1]$ is called a non-additive generalized directed divergence function of order $\alpha (\alpha \neq 1)$ if h is a solution of the functional equation,

$(2.4.7) \quad h(x,y,z) + (1-x)(1-y)^{\alpha-1}(1-z)^{1-\alpha}\, h(u/(1-x), v/(1-y), w/(1-z))$

$\qquad = h(u,v,w) + (1-u)(1-v)^{\alpha-1}(1-w)^{1-\alpha}\, h(x/(1-u), y/(1-v), z/(1-w))$

for $x,y,z,u,v,w \in [0,1[,\ x+u,\ y+v,\ z+w \in [0,1]$ satisfying further

$(2.4.8) \quad h(0,0,0) = h(1,1,1)$

and

$(2.4.9) \quad h(1,1,\tfrac{1}{2}) = h(0,0,\tfrac{1}{2}) = 1$

Definition 2.4.5: If h is a non-additive generalized directed divergence function of order $\alpha(\alpha \neq 1)$ as defined above, then the non-additive generalized directed divergence of order $\alpha(\alpha \neq 1)$ is defined by the relation

$(2.4.10) \quad I_{n,\alpha}(P:Q:R) = \sum_{i=2}^{n} P_i Q_i^{\alpha-1} R_i^{1-\alpha}\, h(p_1/P_i, q_i/Q_i, r_1/R_1)$

where $P_i = p_1 + \ldots + p_i$, $Q_i = q_1 + \ldots + q_i$ and $R_i = r_1 + \ldots + r_i$, for $i = 1,\ldots,n$. Generalized measure of directed divergence given in definition 2.4.1 has applications in various disciplines. For example, an interpretation of it as a measure of 'information improvement' of forecast revision in Econometrics may be seen from Theil (1967, p.43).

(b) PROPERTIES OF (2.4.1) TO (2.4.5):

Most of the properties of the various measures defined in this section are similar to what we have stated in the earlier section of this chapter.

(i) Functional Equation: Let $f(p,q,r) = I_2(p, 1-p: q, 1-q: r, 1-r)$. Then f satisfies the functional equation,

(2.4.11) $\quad f(x,y,z) + (1-x) f(u/(1-x), v/(1-y), w/(1-z))$

$$= f(u,v,w) + (1-u) f(x/(1-u), y/(1-v), z/(1-w))$$

for $x, y, z, u, v, w \in [0,1[$ with $x+u, y+v, z+w \in [0,1]$.

(ii) All the measures satisfy *symmetry*, *expansibility* and *continuity* properties discussed for other measures earlier. Also (2.4.5) satisfies additivity.

(iii) Recursivity: For $P, Q, R \in S_n$ and $p_1+p_2, q_1+q_2, r_1+r_2 > 0$,

(2.4.12) $\quad I_{n,\alpha}(p_1, \ldots, p_n : q_1, \ldots, q_n : r_1, \ldots, r_n)$

$$= I_{n-1,\alpha}(p_1+p_2, p_3, \ldots, p_n : q_1+q_2, q_3, \ldots, q_n : r_1+r_2, r_3, \ldots, r_n)$$

$$+ (p_1+p_2)(q_1+q_2)^{\alpha-1}(r_1+r_2)^{1-\alpha} I_{2,\alpha}(p_1/(p_1+p_2), p_2/(p_1+p_2) :$$

$$q_1/(q_1+q_2), q_2/(q_1+q_2) : r_1/(r_1+r_2), r_2/(r_1+r_2))$$

(iv) Non-additivity: For $P, Q, R \in S_n$, $p'_j, q'_j, r'_j \geq 0$ with $\sum_{j=1}^{m} p'_j = \sum_{j=1}^{m} q'_j = \sum_{j=1}^{m} r'_j = 1$,

(2.4.13) $\quad I_{mn,\alpha}(p_1 p'_1, \ldots, p_1 p'_m, \ldots, p_n p'_1, \ldots, p_n p'_m : q_1 q'_1, \ldots, q_1 q'_m, \ldots, q_n q'_1,$

$$\ldots, q_n q'_m : r_1 r'_1, \ldots, r_1 r'_m, \ldots, r_n r'_1, \ldots, r_n r'_m)$$

$$= I_{n,\alpha}(p_1, \ldots, p_n : q_1, \ldots, q_n : r_1, \ldots, r_n) + I_{m,\alpha}(p'_1, \ldots, p'_m : q'_1, \ldots, q'_m :$$

$$r'_1, \ldots, r'_m) + (2^{\alpha-1}-1) I_{n,\alpha}(p_1, \ldots, p_n : q_1, \ldots, q_n : r_1, \ldots, r_n)$$

$$\times I_{m,\alpha}(p'_1, \ldots, p'_m : q'_1, \ldots, q'_m : r'_1, \ldots, r'_m)$$

(c) GENERAL FUNCTIONAL EQUATION

In this section we solve a general functional equation concerning generalized directed divergence from which characterization theorems can be established. Some of the theorems on generalized directed divergences are put as excercises at the end of this chapter.

For $x, y, z, u, v, w \in [0,1[$, $x+u, y+v, z+w \in [0,1]$ consider the functional equation in $f(x,y,z): [0,1] \times [0,1] \times [0,1] \to \mathbb{R}$,

(2.4.14) $\quad f(x,y,z) + G(x,y,z) f(u/(1-x), v/(1-y), w/(1-z))$

$$= f(u,v,w) + G(u,v,w)\ f(x/(1-u),y/(1-v),z/(1-w))$$

where $G:[0,1] \times [0,1] \times [0,1] \to \mathcal{R}$ is a given function satisfying

(2.4.15) $\quad G(x+u-xu,\ y+v-yv,\ z+w-zw) = G(x,y,z)\ G(u,v,w)$

Assume the following conditions on f

(2.4.16) $\quad f(0,0,0) = f(1,1,1)$

and

(2.4.17) $\quad f(0,0,½) = f(1,1,½) = 1$

The general solution of (2.4.14) under the conditions (2.4.15),(2.4.16) and (2.4.17) is given recently by Kannappan and Rathie (1974). Here we will first discuss the functional equation (2.4.15) and later on solve (2.4.14) in detail for one case which is important from the point of view of information theoretic applications.

Taking $y=v=w=z=0$ in (2.4.15) and substituting $g(x) = G(x,0,0)$, $x \in [0,1]$, we find that g satisfies

(2.4.18) $\quad \Phi(x+y-xy) = \Phi(x)\ \Phi(y)$, for $x,y \in [0,1]$.

Similarly, it is easy to see that $h(y) = G(0,y,0)$, $y \in [0,1]$ and $k(z) = G(0,0,z)$, $z \in [0,1]$ also satisfy (2.4.18). Again, (2.4.15) for $u=y=z=w=0$ takes the following form:

(2.4.19) $\quad G(x,v,0) = G(x,0,0)\ G(0,v,0)$, $x,v \in [0,1]$.

Hence (2.4.15) for $u=v=z=0$ with the help of (2.4.19) gives

(2.4.20) $\quad G(u,v,w) = g(u)\ h(v)\ k(w)$, $u,v,w \in [0,1]$.

Thus (2.4.14) due to (2.4.20) reduces to the following form:

(2.4.21) $\quad f(x,y,z) + g(x)\ h(y)\ k(z)\ f(u/(1-x),v/(1-y),w/(1-z))$

$$= f(u,v,w) + g(u)\ h(v)\ k(w)\ f(x/(1-u),y/(1-v),z/(1-w))$$

for $x,y,z,u,v,w \in [0,1[$ with $x+u,\ y+v,\ z+w \in [0,1]$. The solutions of (2.4.18) are given below.

(i) $\quad \Phi(x) = 0,\ x \in [0,1]$

(ii) $\quad \Phi(x) = 0,\ x \in]0,1],\ \Phi(0) = 1$

(iii) $\quad \Phi(x) = 1,\ x \in [0,1]$

(iv) $\Phi(x) = 1$, $x \in [0,1[$, $\Phi(1) = 0$

(v) $\Phi(x)$ is non-constant for $x \in]0,1[$ with $\Phi(0) = 1$ and $\Phi(1) = 0$.

For more details, see Kannappan and Rathie (1974) and Aczél, Baker, Djoković, Kannappan and Rado (1971). From *(i)* to *(v)* above, it is clear that we will have in all 125 cases to be discussed in order to solve the equation *(2.4.21)* completely. Some of these cases will be trivial and similar in nature.

Now we will solve *(2.4.21)* under the conditions *(2.4.16)* and *(2.4.17)* in only one important case where all the functions g, h and k are given by *(v)*. The result is given in the following theorem.

THEOREM 2.4.1: If g, h and k are non-constant solutions of *(2.4.18)* satisfying the boundary conditions

(2.4.22) $g(0) = h(0) = k(0) = 1$

and

(2.4.23) $g(1) = h(1) = k(1) = 0$

then the general solution of the functional equation *(2.4.21)* under the consitions *(2.4.16)* and *(2.4.17)* is given by

(2.4.24) $f(x,y,z) = [g(x)\, h(y)\, k(z) + g(1-x)\, h(1-y)\, k(1-z) - 1]/[k(\tfrac{1}{2}) - 1]$,
$\qquad x,y,z \in [0,1]$.

Proof: The equation *(2.4.21)* for $p = u/(1-x)$, $q = v/(1-y)$, $r = w/(1-z)$, $\xi = 1-x$, $\eta = 1-y$ and $\zeta = 1-z$ reduces to

(2.4.25). $f(1-\xi, 1-\eta, 1-\zeta) + g(1-\xi)\, h(1-\eta)\, k(1-\zeta)\, f(p,q,r)$

$\qquad = f(p\xi, q\eta, r\zeta) + g(p\xi)\, h(q\eta)\, k(r\zeta)\, f((1-\xi)/(1-p\xi), (1-\eta)/(1-q\eta),$

$\qquad\qquad (1-\zeta)/(1-r\zeta))$, for $p,q,r \in [0,1]$, $\xi,\eta,\zeta \in]0,1]$ with $p\xi \ne 1$,

$\qquad\qquad q\eta \ne 1$, $r\zeta \ne 1$.

Now, for $\xi=1$, $\eta=1$, $\zeta=\tfrac{1}{2}$ and $r=0$, *(2.4.25)* takes the form:

(2.4.26) $f(0,0,\tfrac{1}{2}) + k(\tfrac{1}{2})\, f(p,q,0) = f(p,q,0) + g(p)\, h(q)\, f(0,0,\tfrac{1}{2})$, $p,q \in [0,1[$.

If $k(\tfrac{1}{2}) \ne 1$, then *(2.4.26)* gives

(2.4.27) $f(p,q,0) = [g(p)\, h(q) - 1]/[k(\tfrac{1}{2}) - 1]$, $p,q \in [0,1[$.

If $k(\tfrac{1}{2})=1$ then *(2.4.26)* gives the values of g and h as constants contradicting the assumptions of the theorem and hence the case $k(\tfrac{1}{2})=1$ is not feasible. The equation (2.4.25) for $p=q=1$ and $r=0$ gives

(2.4.28) $\quad f(1-\xi,\ 1-\eta,\ 1-\zeta) + g(1-\xi)\ h(1-\eta)\ k(1-\zeta)\ f(1,1,0)$

$\quad\quad = f(\xi,\ \eta,\ 0) + g(\xi)\ h(\eta)\ f(1,1,1-\zeta)\ ,\quad \xi,\eta \in\ ]0,1[,\ \zeta \in\ ]\ 0,1]\ .$

Now, (2.4.28) with $\zeta=1, \eta=\tfrac{1}{2}$ and (2.4.27) give

(2.4.29) $\quad f(1,1,0) = 1/(1-k(\tfrac{1}{2}))$

Taking $\eta=1$, $\zeta=1$ and $p=0$ in (2.4.25) and utilizing (2.4.27) we have

(2.4.30) $\quad f(0,q,r) = [h(q)\ k(r) - 1]/\ [k(\tfrac{1}{2}) - 1],\quad q,r \in\ [0,1[$

Putting $p=0$, $q=0$ and $r=1$ in (2.4.25), we have

(2.4.31) $\quad f(1-\xi,\ 1-\eta,\ 1-\zeta) + g(1-\xi)\ h(1-\eta)\ k(1-\zeta)\ f(0,0,1)$

$\quad\quad = f(0,0,\zeta) + k(\zeta)\ f(1-\xi,\ 1-\eta,\ 1),\quad \xi,\eta \in\ ]0,1],\ \zeta \in\ ]0,1[$

Now (2.4.31) with $\xi=1$, $\eta=1$ and (2.4.30) with $q=0$ give

(2.4.32) $\quad f(0,0,1) = 1/[1-k(\tfrac{1}{2})]$

The equation (2.4.28) with $\zeta=\tfrac{1}{2}$, on utilizing (2.4.17), (2.4.27) and (2.4.29), gives

(2.4.33) $\quad f(1-\xi,\ 1-\eta,\tfrac{1}{2}) = [k(\tfrac{1}{2})(g(\xi)\ h(\eta) + g(1-\xi)\ h(1-\eta)) -1]/[k(\tfrac{1}{2}) -1]$

for $\xi,\eta \in\ ]0,1[$. Now, from (2.4.31) with $\zeta=\tfrac{1}{2}$ on utilizing (2.4.17), (2.4.32) and (2.4.33), we have

(2.4.34) $\quad f(1-\xi,\ 1-\eta,\ 1) = [g(\xi)\ h(\eta) -1]/[k(\tfrac{1}{2}) - 1],\quad \xi,\eta \in\ ]0,1[$

Again, (2.4.31) on using (2.4.30), (2.4.32) and (2.4.34) gives

(2.4.35) $\quad f(1-\xi,\ 1-\eta,\ 1-\zeta) = [g(\xi)\ h(\eta)\ k(\zeta) + g(1-\xi)\ h(1-\eta)\ k(1-\zeta) -1]/$

$\quad\quad\quad [k(\tfrac{1}{2}) - 1]\ ,\quad \xi,\eta,\zeta \in\ ]0,1[$

Following similar procedures it is not very difficult to obtain $f(1,1,1-\zeta)$, $f(1,0,0)$, $f(1,1-\eta,\ 1-\zeta)$, $f(1,\ 1-\eta,\ 0)$, $f(0,1,1)$, $f(1-\xi,\ 1,1)$, $f(p,0,r)$, $f(1,0,1)$, $f(1,\ 1-\eta,\ 1)$, $f(0,1,0)$, $f(1-\xi,\ 1,\ 1-\zeta)$, $f(1-\xi,\ 1,\ 0)$, $f(1,\ 0,\ 1-\zeta)$ and $f(0,\ 1-\eta,\ 1)$. This proves theorem 2.4.1.

A characterization theorem for a generalized directed divergence can be proved by assuming recursivity, symmetry and normalization. For details, see Kannappan and Rathie (1974).

2.5 PSEUDO-MEASURES OF DIRECTED DIVERGENCES

Consider a finite discrete probability distribution $P = (p_1, \ldots, p_n)$, $p_i \geq 0$, $\Sigma p_i = 1$ and a sequence of numbers $(q_1, \ldots, q_n)$ $q_i \geq 0$ and $\Sigma q_i \leq 1$. In this section we will define a measure based on the p_i's and q_i's and study some of its properties.

(a) *DEFINITION 2.5.1:*

The following quantity,

(2.5.1) $\quad J_n(p_1,\ldots,p_n:q_1,\ldots,q_n) = \Sigma\, p_i\, \log(p_i/q_i)$

with the convention as in section 2.1, is defined as the pseudo-directed divergence. In a similar way the other quantities of the previous section may be defined where $p_i \geq 0$, $i=1,\ldots,n$, $\Sigma p_i = 1$, $q_i \geq 0$, $i=1,\ldots,n$, $\Sigma q_i \leq 1$ and $r_i \geq 0$, $i=1,\ldots,n$, $\Sigma r_i \leq 1$.

(b) *CHARACTERIZATION*

Consider the following postulates:

E1. *Recursivity:* (2.1.6) for $\Sigma p_i = 1$, $\Sigma q_i = 1$ and for all $n \geq 3$
E2. *Symmetry:* (2.1.5) for $n = 3$ where $p_1+p_2+p_3 = 1 = q_1+q_2+q_3$
E3. *Continuity:* Let $f(x,y) = J_2(x,1-x:y,1-y)$, for $(x,y) \in J$, where $J =]0,1[\times]0,1[\, \cup\, \{(0,y)\} \cup \{(1,y')\}$ with $y \in [0,1]$ and $y' \in]0,1]$. Let $f(x,y)$ be continuous on J.
E4. For $\Sigma_{i=1}^{n-1} p_i = 1$, $\Sigma_{i=1}^{n} q_i \leq 1$ for all $n \geq 2$, $J_n(p_1,\ldots,p_{n-1},0:q_1,\ldots,q_n)$
 $= J_{n-1}(p_1,\ldots,p_{n-1}:q_1,\ldots,q_{n-1})$
E5. *Normalization:* $J_2(1,0:\tfrac{1}{2},\tfrac{1}{2}) = 1$
E6. *Nullity:* $J_2(\tfrac{1}{2},\tfrac{1}{2}:\tfrac{1}{2},\tfrac{1}{2}) = 0$.

A theorem due to Rathie and Kannappan (1973a), to be proved later, will depend on the following lemmas.

LEMMA 2.5.1: For $(x,y) \in J$,

(2.5.2) $\quad f(x,y) = f(1-x,1-y)$

The proof of this lemma is the same as given in lemma 2.1.1.

LEMMA 2.5.2: For $x,y,u,v \in [0,1[$, $x+u$, $y+v \in [0,1]$, f satisfies

(2.5.3) $\quad f(x,y) + (1-x)\, f(u/(1-x),v/(1-y)) = f(u,v) + (1-u)\, f(x/(1-u),y/(1-v))$

The proof is the same as given in lemma 2.1.2. Taking $x=y=0$ in

(2.5.3) we get $f(0,0)=0$ which with the help of (2.5.2) implies

(2.5.4) $\quad f(0,0) = f(1,1) = 0$

LEMMA 2.5.3: The function J_n is symmetric in pairs for all n and it is given by

(2.5.5) $\quad J_n(p_1,\ldots,p_n;q_1,\ldots,q_n) = \sum_{i=2}^{n} r_i\, f(p_i/r_i,\, q_i/s_i)$

where $r_i = p_1 + \ldots + p_i$ and $s_i = q_1 + \ldots + q_i$.

Proof: Equation (2.5.5) can be easily established by induction from postulates E1, E3 and lemma 2.5.1.

By lemma 2.5.1 and postulate E2, it is clear that J_2 and J_3 are symmetric. So we have only to show that J_n is symmetric in pairs $\{p_i, q_i\}$ for all $i \geq 4$ with $\Sigma p_i = \Sigma q_i = 1$. Let J_{n-1} be symmetric in pairs $\{p_i, q_i\}$, $i=1,\ldots,n-1$. Then by postulate E1 we find that J_n is symmetric in $\{p_1, q_1\}$, $\{p_2, q_2\}$ and in pairs in $\{p_3, q_3\}$, $\{p_4, q_4\},\ldots, \{p_n, q_n\}$ respectively. In order to prove that J_n is symmetric, it is enough to prove that J_n is symmetric in $\{p_2, q_2\}$ and $\{p_3, q_3\}$. From postulate E1 we get

(2.5.6) $\quad J_n(p_1,p_3,p_2,p_4,\ldots,p_n;q_1,q_3,q_2,q_4,\ldots,q_n) = J_{n-1}(p_1+p_3,p_2,p_4,\ldots,p_n;$

$q_1+q_3,q_2,q_4,\ldots,q_n) + (p_1+p_3)\, J_2(p_1/(p_1+p_3),p_3/(p_1+p_3);q_1/(q_1+q_3),$

$q_3/(q_1+q_3))$

In order to prove lemma 2.5.3 we have to show that the two J_n's given by the postulate E1 and (2.5.6) are equal, that is,

(2.5.7) $\quad J_{n-1}(p_1+p_3,p_2,p_4,\ldots,p_n;q_1+q_3,q_2,q_4,\ldots,q_n) + (p_1+p_3)\, J_2(p_1/(p_1+p_3),$

$p_3/(p_1+p_3);q_1/(q_1+q_3),q_3/(q_1+q_3))$

$= J_{n-1}(p_1+p_2,p_3,p_4,\ldots,p_n;q_1+q_2,q_3,q_4,\ldots,q_n) + (p_1+p_2)\, J_2(p_1/(p_1+p_2),$

$p_2/(p_1+p_2);q_1/(q_1+q_2),q_2/(q_1+q_2))$, for $n \geq 4$

The equation (2.5.7) is true if the following equation, obtained by applying the postulate E1 to (2.5.7), is true.

(2.5.8) $\quad (p_1+p_2+p_3)\, J_2((p_1+p_2)/(p_1+p_2+p_3),p_3/(p_1+p_2+p_3);(q_1+q_2)/(q_1+q_2+q_3),$

$q_3/(q_1+q_2+q_3))+(p_1+p_2)J_2(p_1/(p_1+p_2),p_2/(p_1+p_2);q_1/(q_1+q_2),q_2/(q_1+q_2))$

Pseudo-Measures of Directed Divergences   63

$$= (p_1+p_3) J_2(p_1/(p_1+p_3), p_3/(p_1+p_3) : q_1/(q_1+q_3), q_3/(q_1+q_3)) +$$
$$(p_1+p_2+p_3) J_2((p_1+p_3)/(p_1+p_2+p_3), p_2/(p_1+p_2+p_3) : (q_1+q_3)/(q_1+q_2+q_3),$$
$$q_2/(q_1+q_2+q_3))$$

But *(2.5.8)* is true since it can be easily reduced to *(2.5.3)*. This proves lemma 2.5.3.

LEMMA 2.5.4: For $\sum_{i=1}^{n} p_i = \sum_{i=1}^{n} q_i = 1$, $p_i = \sum_{j=1}^{m_i} p_{ij} > 0$ and $q_i = \sum_{j=1}^{m_i} q_{ij} > 0$,

(2.5.9) $\quad J_{m_1+\ldots+m_n}(p_{11},\ldots,p_{1m_1},\ldots,p_{n1},\ldots,p_{nm_n} : q_{11},\ldots,q_{1m_1},\ldots,q_{n1},$
$$\ldots, q_{nm_n}) = J_n(p_1,\ldots,p_n : q_1,\ldots,q_n)$$
$$+ \sum_{i=1}^{n} p_i J_{m_i}(p_{i1}/p_i,\ldots,p_{im_i}/p_i : q_{i1}/q_i,\ldots,q_{im_i}/q_i)$$

Proof: Applying successively postulate *E1* to the first m_1 pairs $\{p_{1i}, q_{1i}\}$ of the *L.H.S.* of *(2.5.9)* one gets

(2.5.10) $\quad$ L.H.S. of (2.5.9) $= J_{1+m_2+\ldots+m_n}(p_1, p_{21},\ldots,p_{2m_n},\ldots,p_{n1},\ldots,p_{nm_n},$
$$q_1, q_{21},\ldots,q_{2m_n},\ldots,q_{n1},\ldots,q_{nm_n}) + \sum_{i=2}^{m_1} r_{1i} J_2(r_{1(i-1)}/r_{1i},$$
$$p_{1i}/r_{1i} : s_{1(i-1)}/s_{1i}, q_{1i}/s_{1i})$$

where $r_{1i} = p_{11}+p_{12}+\ldots+p_{1i}$, $p_1 = r_{1m_1}$, $s_{1i} = q_{11}+q_{12}+\ldots+q_{1i}$, $q_1 = s_{1m_1}$.

Equation *(2.5.10)* on utilizing lemma *2.5.3* yields

(2.5.11) $\quad$ L.H.S. of (2.5.9) $= J_{1+m_2+\ldots+m_n}(p_1, p_{21},\ldots,p_{2m_2},\ldots,p_{n1},\ldots,p_{nm_n} :$
$$q_1, q_{21},\ldots,q_{2m_2},\ldots,q_{n1},\ldots,q_{nm_n}) + p_1 J_{m_1}(p_{11}/p_1, p_{12}/p_1,\ldots$$
$$\ldots,p_{1m_1}/p_1 : q_{11}/q_1, q_{12}/q_1,\ldots,q_{1m_1}/q_1)$$

In a similar way, again applying postulate *E1* successively to the first term on the right hand side of *(2.5.11)* we arrive at *(2.5.9)*. In this process we have made use of the symmetry proved in lemma *2.5.3*.

If $p_i=0$ for any i in the lemma *2.5.4* first we use postulate *E4* and then apply lemma *2.5.4*. In other words, if a $p_i=0$, then in *(2.5.9)* the corresponding term in the sum on the right side is assumed to be zero. A special case of lemma *2.5.4* for $m_1=\ldots=m_n=m$, which will be used later on, is given below.

(2.5.12) $J_{nm}(p_{11},\ldots,p_{1m},\ldots,p_{n1},\ldots,p_{nm}:q_{11},\ldots,q_{1m},\ldots,q_{n1},\ldots,q_{nm})$
$= J_n(p_1,\ldots,p_n:q_1,\ldots,q_n) + \Sigma_{i=1}^n p_i J_m(p_{i1}/p_i,\ldots,p_{im}/p_i:q_{i1}/q_i,\ldots,q_{im}/q_i)$

Let

(2.5.13) $\Phi(m,n) = J_n(1/m,\ldots,1/m,0,\ldots,0:1/n,\ldots,1/n)$, for $n \geq m$

Next we prove a lemma for $\Phi(m,n)$.

LEMMA 2.5.5: For $1 \leq t \leq n$, $1 \leq s \leq m$ where t,n,s and m are integers,
(2.5.14) $\Phi(st,mn) = \Phi(s,m) + \Phi(t,n)$

Proof: Substituting $p_{ij}=1/ts$ for $i=1,\ldots,t$, $j=1,\ldots,s$ and $p_{ij}=0$ otherwise and $q_{ij}=1/mn$ for all $i=1,\ldots,n$, $j=1,\ldots,m$ in (2.5.12) we get

(2.5.15) $J_{mn}(\underbrace{1/ts,\ldots,1/ts}_{mt\text{ terms}},0,\ldots,0,\ldots,1/ts,\ldots,1/ts,0,\ldots,0,0,\ldots,0,\ldots 0:$
$1/nm,\ldots,1/nm) = J_n(1/t,\ldots,1/t,0,\ldots,0:1/n,\ldots,1/n)$
$+ J_m(1/s,\ldots,1/s,0,\ldots,0:1/m,\ldots,1/m)$

The equation (2.5.15) on using (2.5.13) and the lemma 2.5.3 proves the lemma 2.5.5.

LEMMA 2.5.6: For $n \leq r$,
(2.5.16) $\Phi(n,r) = \log(r/n)$

Proof: The equation (2.5.14) for $s=t=1$ gives

(2.5.17) $\Phi(1,nm) = \Phi(1,m) + \Phi(1,n)$, for $m \geq 1$, $n \geq 1$

From (2.5.13) and (2.5.5), one obtains

(2.5.18) $\Phi(1,r) = J_r(1,0,\ldots,0:1/r,\ldots,1/r) = \Sigma_{i=2}^r f(0,1/i)$

Therefore, we get

(2.5.19) $\Phi(1,r+1) - \Phi(1,r) = f(0,1/(r+1))$

Hence from (2.5.19) and (2.5.4) we find that

(2.5.20) $\lim_{r \to \infty}[\Phi(1,r+1) - \Phi(1,r)] = f(0,0) = 0$

Pseudo-Measures of Directed Divergences

Thus the solution of the functional equation *(2.5.17)* under the condition *(2.5.20)* as given by Erdös (1946) is as follows.

(2.5.21) $\Phi(1,r) = b \log r$, $r \geq 1$

where b is an arbitrary real constant. For $t=n$ and $s=m$ in *(2.5.14)* we get

(2.5.22) $\Phi(mn,mn) = \Phi(m,m) + \Phi(n,n)$

By *(2.5.13)* and *(2.5.5)* we get

(2.5.23) $\Phi(n,n) = (1/n) \sum_{i=2}^{n} i \, f(1/i,1/i)$

Now due to *(2.5.4)* and $\lim_{n \to \infty} \alpha_i F_i = 0$ if $\sum_{i=1}^{n} \alpha_i = 1$ and $\lim_{i \to \infty} F_i = 0$, the equation *(2.5.23)* yields

(2.5.24) $\lim_{n \to \infty} [\Phi(n+1,n+1) - \Phi(n,n)] = \lim_{n \to \infty} [f(1/(n+1), 1/(n+1))$
$$- 2^{-1} \sum_{i=1}^{n} \{2i/n(n+1)\} \, f(1/i,1/i)] = 0$$

Hence the equations *(2.5.22)* and *(2.5.24)*, and Erdös(1946) give

(2.5.25) $\Phi(n,n) = c \log n$, for $n \geq 1$

Now the postulate *E6* and *(2.5.25)* give

(2.5.26) $\Phi(n,n) = 0$

Hence from *(2.5.14)* for $t=n$, $s=1$, *(2.5.21)* and *(2.5.26)*, we have

(2.5.27) $\Phi(n,mn) = b \log m$, for $m \geq 1$

Also *(2.5.24)* for $t=1$ and $m \geq n$ gives

(2.5.28) $\Phi(n,mn) = \Phi(n,m) + \Phi(1,n)$

Hence *(2.5.28)*, *(2.5.27)*, *(2.5.21)* and postulate *E5* prove lemma 2.5.6.

LEMMA 2.5.7: For all $(x,y) \in J$

(2.5.29) $f(x,y) = x \log(x/y) + (1-x) \log\{(1-x)/(1-y)\}$

Proof: Substituting $n = 2$, $m_1 = pr$, $m_2 = qr-pr$, $p_{11} = \ldots = p_{1m} = 1/s$, $p_{1(m+1)} = \ldots = p_{1(pr)} = 0$, $q_{11} = \ldots = q_{1(pr)} = 1/qr$, $p_{21} = \ldots = p_{2(s-m)} = 1/s$, $p_{2(s-m+1)} = \ldots = p_{2(qr-pr)} = 0$, $q_{21} = \ldots = q_{2(qr-pr)} = 1/qr$ in *(2.5.9)*, one gets

(2.5.30) $\quad J_{qr}(1/s,\ldots,1/s,0,\ldots,0,1/s,\ldots,1/s,0,\ldots,0:1/(qr),\ldots,1/(qr)\ )$

$\qquad = J_2(m/s,1-m/s:pr/(qr),1-pr/(qr),1-pr/(qr)) + (m/s)\ J_{pr}(1/m,\ldots,1/m,$

$\qquad 0,\ldots,0:1/(pr),\ldots,1/(pr)\ ) + (1-m/s)\ J_{qr-pr}(1/(s-m),\ldots,1/(s-m),$

$\qquad 0,\ldots,0:1/(qr-pr),\ldots,1/(qr-pr),\ldots,1/(qr-pr)\ )$

for $pr \geqslant m$, $(q-p)r \geqslant s-m > 0$ and m,p,q,r,s are positive integers. Thus (2.5.30), $f(x,y)$ in postulate E3, (3.5.13) and lemma 2.5.3 yield,

(2.5.31) $\quad f(m/s,p/q) = \Phi(s,qr) - (m/s)\ \Phi(m,pr) - (1-m/s)\ \Phi(s-m,qr-pr)$

for $pr \geqslant m$ and $(q-p)r \geqslant s-m > 0$. Hence from (2.5.31) and (2.5.16) we have

(2.5.32) $\quad f(m/s,p/q) = (m/s)\ log\{(m/s)/(p/q)\} + (1-m/s)\ log\{(1-m/s)(1-p/q)\}$

for $s > m$, $q > p$. Equation (2.5.32) gives $f(x,y)$ for all rationals $x,y \in\ ]0,1[$. Now consider any point (x,y) with $x,y \in\ ]0,1[$ and two sequences of rational numbers $\{r_n\}$ and $\{s_n\}$ such that $x < r_n \leqslant 1$ and $y < s_n \leqslant 1$ with $\lim\limits_{n\to\infty} r_n = x$ and $\lim\limits_{n\to\infty} s_n = y$. Let $x_n = 1-x/r_n$ and $y_n = 1-y/s_n$ so that $x_n, y_n \in\ ]0,1[$. Now taking $u = x_n$, $v = y_n$ in (2.5.3), we have

(2.5.33) $\quad f(x,y) + (1-x)\ f(x_n/(1-x), y_n/(1-y)) - f(x_n, y_n)$

$\qquad = (1-x_n)\ f(x/(1-x_n), y/(1-y_n))$

By virtue of postulate E3, limit, as $x_n \to 0$, $y_n \to 0$ (due to $n \to \infty$), of the left side of (2.5.33) exists. Hence taking limit as $n \to \infty$ in (2.5.33) and using (2.5.4) and (2.5.32), we get

(2.5.34) $\quad f(x,y) = \lim\limits_{n\to\infty} (1-x_n)\ f(x/(1-x_n), y/(1-y_n))$

$\qquad = \lim\limits_{n\to\infty} (x/r_n)\ f(r_n, s_n)$

$\qquad = \lim\limits_{n\to\infty} (x/r_n)[r_n\ log(r_n/s_n) + (1-r_n)\ log\{(1-r_n)/(1-s_n)\}]$

$\qquad = x\ log(x/y) + (1-x)\ log\{(1-x)/(1-y)\}$, for $x,y \in\ ]0,1[$

Now, we have to determine $f(0,y)$ and $f(1,y)$ for $y \in\ ]0,1[$. Taking $x=0$ and $v=1-2y$ in (2.5.3) and using (2.5.34) and postulate E5 we have

(2.5.35) $\quad f(0,y) = -\ log(1-y)$, for $y \in\ ]0,\tfrac{1}{2}[$

Again putting $x=0$, $v=\tfrac{1}{2}$ in (2.5.3) and utilizing (2.5.34) and (2.5.35) we get

Pseudo-Measures of Directed Divergences

(2.5.36) $f(0,y) = -\log(1-y)$, $y \in]0,1[$

Now (2.5.36) and (2.5.2) give

(2.5.37) $f(1,y) = -\log y$, $y \in]0,1[$

Hence (2.5.34), (2.5.36) and (2.5.4) prove lemma 2.5.7 completely.

THEOREM 2.5.1: The function $J_n(p_1,\ldots,p_n;q_1,\ldots,q_n)$, where $P \in S_n$ and $q_i \geq 0$, $\sum_{i=1}^{n} q_i \leq 1$ satisfying the postulates E_1 to E_6 is the pseudo-directed divergence given by (2.5.1).

The proof follows easily from the lemmas and hence it is omitted.

Remark: As pointed out by Rathie and Kannappan (1973a), a characterization theorem for J_n due to Hobson (1969) is incorrect since his result (19) p.388 is not true if $q_i = 0$ which is used in the proof of his theorem.

2.6 CONTINUOUS ANALOGUES

The continuous analogue to the directed divergence, for the two probability measures P and Q with density functions $p(x)$ and $q(x)$ respectively, is defined by

(2.6.1) $I(P:Q) = \int_{-\infty}^{\infty} p(x) \log [p(x)/q(x)] \, dx$

with the usual convention that the integrant is zero whenever $p(x) = 0$.

In a similar manner, continuous analogues corresponding to the various measures defined in this chapter for discrete distributions can be defined and studied.

Various results on directed divergence (2.6.1) can be seen in Kullback (1959), Campbell (1970), Dutta (1966) and Jaynes (1957). Results on f-divergence can be seen in Csiszár (1967, 1967a).

2.7 APPLICATIONS

(a) DIRECTED DIVERGENCE (2.1.1)

As discussed earlier in chapter 1, $-\log p_i$ can be interpreted as the information content in the event E_i with probability p_i. Thus $\log(p_i/q_i) = \log p_i - \log q_i$ may be taken as the *information gain* in predicting the event E_i. For example consider the problem of weather forecast and consider the prediction of the event E. Let the unconditional probability of E, which may be called climatological chance in the case of weather forecast, be q_0. From the data available we may estimate the conditional probability p_0 of E given that E is predicted to occur. Then $log(p_0/q_0) = h(q_0) - h(p_0)$ is usually called the *information gain* and $h(p_0)$ and $h(q_0)$ are called the *information contents* of the weather forecast respectively. For other similar interpretations in Econometric problems see Theil (1967, pp.8-20). The average information gain, that is, $\Sigma p_i \, log(p_i/q_i)$ is often interpreted as the *mean information for discrimination in favour of hypothesis H_1 against hypothesis H_2* in statistical inference problems (Kullback (1959)). Kerridge (1961) interprets $\Sigma p_i \, log(p_i/q_i)$ as a *measure of the error* made by the observer in estimating a discrete probability distribution as $Q = (q_1,\ldots,q_n)$ which is in fact $P = (p_1, \ldots, p_n)$.

The directed divergence *(2.1.1)* is also interpreted as the *information content of the increase forecast* in survey forecasts of economic commodities based on prediction-realization tables for purchase prices (Theil ,1967 ,p.27). It can also be interpreted as the *information inaccuracy* of the forecast $(q_1,\ldots,q_n)$ with respect to the realizations $(p_1,\ldots,p_n)$. An application of information inaccuracy to construction industry, input structure prediction etc., may be seen from Theil (1967, pp.36-42). The measure *(2.1.1)* may be looked upon as the logarithm of a weighted geometric mean *deflated* per capita incomes among groups of individuals (Theil, 1967,pp.102-105).

Consider the problem in Psychology where we are defining a measure of *psychological information content* of the event i. Let q_i be the *psychological probability* of the event i and let p_i be the real probability. Hence a measure of the amount of information, which one believes one gets, may be taken as $-\Sigma p_i \, log \, q_i$. Hence

$$-\Sigma p_i \, log \, q_i - (-\Sigma p_i \, log \, p_i) = \Sigma p_i \, log(p_i/q_i)$$

may be interpreted as the *excess apparent information* which is the consequence of incorrect estimation of probabilities. Also *(2.1.1)* was interpreted as *information radius* by Sibson (1969).

Directed divergence defined in *(2.1.1)* has been known in the literature for a long time - see for example, Kullback (1959), Quastler (1956) and Fano (1949).

(b) *GENERALIZED DIRECTED DIVERGENCE* (2.4.1)

We have pointed out earlier that $\Sigma p_i \, log(p_i/q_i)$ may be interpreted as information inaccuracy of the input structure. Suppose

Applications

that more data are available now, then we can make a prediction revision. Let the revised forecast of p_i's be denoted by r_i's compared with the original forecast q_i's. Then $\Sigma p_i \log(p_i/q_i) - \Sigma p_i \log(p_i/r_i) = \Sigma p_i \log(r_i/q_i)$ may be interpreted as *information improvement* of the forecast revision (Theil, 1967, p.43).

(c) PSEUDO-MEASURES

The pseudo-measures also have some practical applications. For example, in Coding Theory, for codes to be uniquely decipherable the code lengths m_i $(i=1,\ldots,n)$ should satisfy Kraft's inequality, namely, $\Sigma 2^{-m_i} \leq 1$. If $q_i = 2^{-m_i}$, where m_i are integers, we get the conditions $q_i \geq 0$ and $\Sigma q_i \leq 1$. Thus in some situations we may have to consider pseudo-measures corresponding to the various measures discussed so far. Some such applications in different disciplines can be seen from Ash (1965).

2.8 OPEN PROBLEMS

There are several unsolved problems, some of which are mentioned below.

2.1 Find the general solution of

$$\Sigma g(p_i) f(p_i) + \Sigma g(q_i) f(q_i) \geq \Sigma g(p_i) f(q_i) + \Sigma g(q_i) f(p_i)$$

where g is a given function, $(p_1,\ldots,p_n)$ and $(q_1,\ldots,q_n)$ are two probability distributions. Discuss the case $g(p_i) = p_i$.

2.2 Find the general solution of the functional inequality

$$\Sigma\, g(p_i)\, f(p_i) / f(q_i) \leq 1$$

where g is a given function, $p_i > 0$, $\Sigma p_i = 1$ and $q_i > 0$, $\Sigma q_i = 1$. Also discuss the cases when $g(p_i) = p_i$ and $g(p_i) = p_i^{\beta}$ respectively. (The case $g(p_i) = p_i^{\beta_i}$ under certain conditions was discussed in (Rathie, 1973)).

2.3 Find the general solution of the functional equation

$$f(x,y) + G(x,y)\, g[u/(1-x), v/(1-y)] = h(x,y) + H(x,y)\, k[x/(1-y), y/(1-v)]$$

for $x,y,u,v \in [0,1[$ with $x+u, y+v \in [0,1]$, where G and H are known functions. Extend the same to three variable case.

2.4 Find the measurable solutions of the functional equation

$$F(x,y,z) + (1-x)^{\alpha}(1-y)^{\beta}(1-z)^{\gamma} F[u/(1-x), v/(1-y), w/(1-z)]$$
$$= F(u,v,w) + (1-u)^{\alpha}(1-v)^{\beta}(1-z)^{\gamma} F[x/(1-u), y/(1-v), z/(1-w)]$$

for $x,y,z,u,v,w \in [0,1[$ with $x+u, y+v, z+w \in [0,1]$.

2.5   Derive characterization theorems for *(2.3.1)*, *(2.3.2)*, *(2.3.3)*, *(2.3.4)*, *(2.3.5)* and *(2.3.6)* by assuming different sets of postulates for each of these measures.

2.6   The *J*-divergence between two directed divergences is defined by the expression

$$J(P:Q) = \Sigma\, p_i\, \log(p_i/q_i) + \Sigma\, q_i\, \log(q_i/p_i)$$

Characterize *J(P:Q)* by assuming different sets of postulates.

EXERCISES

2.1   (Campbell, 1972). Discover the class of functions *r* which will satisfy the following conditions.

(1) For each positive integer m, $r(p_1,\ldots,p_m:q_1,\ldots,q_m)$ is a continuous function of the $2m$ variables $p_1,\ldots,q_m$ on the subspace of $\mathcal{R}^{2m}$ defined by the relations $p_i \geq 0$, $q_i > 0$, $\Sigma q_i = 1$;

(2) Symmetric as in property *(iii)* of §2.1;

(3) $r(p_1,\ldots,p_{m-1},0:q_1,\ldots,q_m) = r(p_1,\ldots,p_{m-1}:q_1,\ldots,q_{m-1})$;

(4) Strongly additive as in section *2.1*, property *(vi)* for $\Sigma_{i=1}^{n} \Sigma_{j=1}^{m} p_{ij} = 1$, $p_{ij} \geq 0$, $p_j = \Sigma_{i=1}^{n} p_{ij}$, $q_{ij} > 0$, $q_j = \Sigma_{i=1}^{n} q_{ij}$;

(5) $r(p_1,\ldots,p_m:q_1,\ldots,q_m)$ is a non-decreasing function of each q_i;

(6) $r(1,q) = r(1/m,\ldots,1/m:q/m,\ldots,q/m)$.

2.2   (Kannappan & Rathie, 1974). Let *f* be a real valued function on $I \times I \times I$ where $I = [0,1]$ satisfying the functional equation,

$$f(x,y,z) + (1-x)^{\alpha}(1-y)^{\beta-1}(1-z)^{1-\beta}\, f(u/(1-x), v/(1-y), w/(1-z))$$
$$= f(u,v,w) + (1-u)^{\alpha}(1-v)^{\beta-1}(1-w)^{1-\beta}\, f(x/(1-u), y/(1-v), z/(1-w))$$

for $x,y,z,u,v,w \in [0,1[$, with $x+u$, $y+v$, $z+w \in I$ and the boundary conditions, $f(0,0,0) = f(1,1,1)$ and $f(\tfrac{1}{2},\tfrac{1}{2},\tfrac{1}{2}) = (2^{1-\alpha}-1)/(2^{\beta-1}-1)$. Show that $f(x,y,z)$ is uniquely given by $f(x,y,z) = (x^{\alpha} y^{\beta-1} z^{1-\beta} + (1-x)^{\alpha}(1-y)^{\beta-1}(1-z)^{1-\beta} - 1)/(2^{\beta-1}-1)$, for all $x,y,z \in I$.

2.3   (Kannappan & Rathie, 1973). Prove that the following postulates uniquely determine $J_n(p_1,\ldots,p_n:q_1,\ldots,q_n)$ for $p_i \geq 0$, $\Sigma p_i = 1$, $q_i \geq 0$, $\Sigma q_i \leq 1$ as given in *(2.5.1)* with the convention of section *2.1*.

(1) Recursivity: section *2.1* property *(iv)*;

(2) Symmetry: section *2.1* property *(iii)* for $n = 3$ only;

(3) Let $\emptyset(p,q) = J_2(p,1-p:q,1-q)$ be continuous for $(p,q) \in J = \,]0,1[\times]0,1[\cup\{(0,y)\}\cup\{(1,y')\}$ with $y' \in \,]0,1]$ and

Exercises

$y \in [0,1[$;

(4) $J_n(p_1,\ldots,p_{n-1},0:q_1,\ldots,q_n) = J_{n-1}(p_1,\ldots,p_{n-1}:q_1,\ldots,q_{n-1})$ with $\Sigma_{i=1}^{n-1} p_i = 1$, $\Sigma_{i=1}^{n} q_i \leqslant 1$ for all $n \geqslant 3$;

(5) $J_2(1,0:\frac{1}{2},\frac{1}{2}) = 1$;

(6) $J_2(\frac{1}{2},\frac{1}{2}:\frac{1}{2},\frac{1}{2}) = 0$.

2.4 (Kannappan & Rathie, 1973). Show that pseudo-directed divergence J_n is uniquely determined as (2.5.1) by the following set of postulates.

(1) $J_n(p_1,\ldots,p_n:q_1,\ldots,q_n) = \Sigma_{i=1}^{n} f(p_i,q_i)$, for all $n \geqslant 2$ with $\Sigma p_i = 1 = \Sigma q_i$ where f is a real valued continuous function defined on J;

(2) $J_{2n}(p_1,p,p_1(1-p),\ldots,p_n p,p_n(1-p):q_1 q,q_1(1-q),\ldots,q_n q,q_n(1-q))$
$= J_n(p_1,\ldots,p_n:q_1,\ldots,q_n) + J_2(p,1-p:q,1-q)$, with $p,q \in [0,1]$ and $\Sigma p_i = 1 = \Sigma q_i$;

(3) Problem 2.3 postulates (4), (5) and (6).

2.5 (Kannappan & Rathie, 1973). Show that the following set of postulates uniquely determine $J_n(p_1,\ldots,p_n:q_1,\ldots,q_n)$ as (2.5.1) with the usual convention.

(1) $J_n(p_1,\ldots,p_n:q_1,\ldots,q_n) - J_{n-1}(p_1+p_2,p_3,\ldots,p_n:q_1+q_2,q_3,\ldots,q_n)$
$= \Delta_{n-1}(p_1,p_2:q_1,q_2)$, for all $n \geqslant 3$ with $\Sigma q_i = 1 = \Sigma p_i$ for some function Δ and the function $\Delta_2(p_1,p_2:q_1,q_2)$ is continuous in $D = \{(p_1,p_2:q_1,q_2), p_1,p_2,q_1,q_2 \geqslant 0, p_1+p_2, q_1+q_2 \leqslant 1\}$;

(2) $J_4(p_1,\ldots,p_4:q_1,\ldots,q_4)$ is symmetric in $\{p_i,q_i\}$, $i=1,\ldots,4$ with $\Sigma_{i=1}^{4} p_i = 1 = \Sigma_{i=1}^{4} q_i$;

(3) Problem 2.4 postulates (2) for $n=3$ and $n=4$;

(4) Problem 2.3 postulates (4), (5) and (6).

2.6 (Rathie, 1973). Show that $I_n(p_1,\ldots,p_n:q_1,\ldots,q_n)$ for $n \geqslant 3$, p_i, $q_i > 0$, $\Sigma p_i = 1 = \Sigma q_i$ is uniquely determined as (2.2.1) for $0 < \alpha < 1$ up to a multiplicative constant by the following set of postulates.

(1) $I_n = c \log(\Sigma_{i=1}^{n} p_i f(p_i)/f(q_i))$ for some function f, $c < 0$;

(2) $I_n \geqslant 0$;

(3) $f(.)$ is differentiable in $]0,1[$.

2.7 (Ng, 1974). Let $P=(p_1,\ldots,p_n)$ and $Q=(q_1,\ldots,q_n)$ with $p_i,q_i \geqslant 0$ and $\Sigma p_i = \Sigma q_i = 1$ be finite discrete probability distributions of arbitrary length. Assume (A) Symmetry:(2.1.5), (B) Expansibility: (2.1.4) and (C) Branching: $I_n(p_1,\ldots,p_n:q_1,\ldots,q_n) = I_{n-1}(p_1+p_2,p_3,\ldots,p_n:q_1+q_2,q_3,\ldots,q_n) + \Delta_n(p_1,p_2:q_1,q_2)$ for $n \geqslant 2$. Prove the following result. If a mapping I_n of all pairs (P,Q) of finite discrete probability distributions into the reals $\mathcal{R}$ satisfies (A), (B) and (C), then there exists a mapping $f = [0,1] \times [0,1] \to \mathcal{R}$ with $f(0,0) = 0$ such that I_n is represented by $I_n(p_1,\ldots,p_n;q_1,\ldots,q_n) = I_1(1:1) - f(1,1) + \sum_{i=1}^{n} f(p_i,q_i)$ for all (P,Q). Conversely, if I_n can be so represented by an f with $f(0,0)=0$, then I_n fulfils (A), (B) and (C). (This result can be extended to mappings defined for m-tuples of finite discrete probability distributions).

2.8 (Kannappan & Ng, 1973). Show that the measurable solutions of the functional equation $F(x,y) + (1-x) F(u/(1-x),v/(1-y))$ $= F(u,v) + (1-u) F(x/(1-u),y/(1-v))$, for $x,y,u,v \in [0,1[$ with $x+u$, $y+v \in [0,1]$ are given by $F(x,y) = A[-x \log x - (1-x)\log(1-x)] + B[x \log y + (1-x)\log(1-y)] + Cx$ where A, B and C are arbitrary constants.

2.9 (Kannappan & Ng, 1974). Let $F:[0,1] \times]0,1[\times]0,1[$ be a real valued function, Lebesgue measurable in each of its three variables, satisfying the functional equation $F(x,y,z) + (1-x) F(u/(1-x),v/(1-y),w/(1-z)) = F(u,v,w) + (1-u) F(x/(1-u),y/(1-v),z/(1-w))$, for all $x,u \in [0,1[$, $y,z,v,w \in]0,1[$ with $x+u \in [0,1]$, $y+v$ and $z+w \in]0,1[$. Then show that F is given by $F(x,y,z) = AS(x) + d_1 x \log(y/(1-y)) + e_1 x \log(z/(1-z)) + e_4 x + d_1 \log(1-y) + e_1 \log(1-z)$, where A, d_1, e_1 and e_4 are constants and S is the Shannon function given by $S(x) = -x \log x - (1-x) \log(1-x)$. The converse is also true.

2.10 (Kannappan & Rathie, 1973b). Prove that the only function $I_n(p_1,\ldots,p_n;q_1,\ldots,q_n;r_1,\ldots,r_n)$ satisfying the properties (1) Recursivity: (iii); (2)Symmetry:$(\;(ii)$ for $n = 3)$;(3) Derivative: Let $f(p,q,r) = I_2(p,1-p;q,1-q;r,1-r)$, for all $(p,q,r) \in]0,1[\times]0,1[\times]0,1[\cup\{(0,y,z)\} \cup\{(1,y',z')\}$ with $y,z \in [0,1[$ and $y',z' \in]0,1]$ have continuous first partial derivatives with respect to all the three variables $p,q,r \in]0,1[$; (4) Nullity: $f(p,p,p) = 0$ for $p \;]0,1[$ and $f(2/3,1/3,1/3) = 0$; (5) Normalization: $f(2/3,2/3,1/3) = 1/3$; is the generalized directed divergence given by $(2.4.1)$.

2.11 (Kannappan, 1972). Prove that the most general continuous solution F of the functional equation $\sum_{i=1}^{m} \sum_{j=1}^{n} F(x_i y_j, u_i v_j) = \sum_{i=1}^{m} F(x_i, u_i) + \sum_{j=1}^{n} F(y_j, v_j)$, for x_i, u_i, y_j, $v_j \geq 0$, $\sum_{i=1}^{m} x_i = 1 = \sum_{j=1}^{n} y_j$, $\sum_{i=1}^{m} u_i \leq 1$ and $\sum_{j=1}^{n} v_j \leq 1$ satisfying further $F(1,\frac{1}{2}) = 1$ and $F(\frac{1}{2},\frac{1}{2}) = 0$ is given by $F(x,y) = x \log(x/y)$, for $(x,y) \in$

Exercises

$]0,1[\times]0,1[\cup\{(0,y)\}\cup\{(1,y')\}$ with $y \in [0,1[$ and $y' \in]0,1]$.

2.12 (Kannappan, 1972a). If F is a Lebesgue measurable solution of the functional equation of exercise 2.11 given above for $m=2$, $n=3$ satisfying the conditions $F(\frac{1}{2},\frac{1}{2}) = 0$ and $F(1,1) = 0$, then prove that F is given by $F(x,y) = k \, x \, \log(x/y)$, where k is an arbitrary constant.

2.13 (Kannappan, 1973). Show that the most general Lebesgue measurable solution F of the functional equation $\sum_{i=1}^{2} \sum_{j=1}^{3} F(x_i p_j, y_i q_j, z_i r_j) = \sum_{i=1}^{2} F(x_i, y_i, z_i) + \sum_{j=1}^{3} F(p_j, q_j, r_j)$ for $x_i, y_i, z_i, p_j, q_j, r_j \geq 0$, $\sum_{i=1}^{2} x_i = 1 = \sum_{j=1}^{3} p_j$, $\sum_{i=1}^{2} y_i \leq 1$, $\sum_{i=1}^{2} z_i \leq 1$, $\sum_{j=1}^{3} q_j \leq 1$, $\sum_{j=1}^{3} r_j \leq 1$ satisfying $F(1,1,1) = F(\frac{1}{2},\frac{1}{2},\frac{1}{2}) = F(1,\frac{1}{2},\frac{1}{2}) = 0$ has the form $F(x,y,z) = k \, x \, \log(y/z)$ on $]0,1[\times]0,1[\times]0,1[\cup\{(0,y,z)\}\cup\{(1,y,z)\}\cup\{(1,1,z')\}\cup\{(0,y',0)\}\cup\{(0,0,z)\}\cup\{(1,y,1)\}$, with $y,z \in]0,1[$, $y' \in [0,1[$, $z' \in]0,1]$, where k is an arbitrary constant and conversely.

2.14 (Kannappan and Rathie, 1972). If h is a solution of the functional equation (2.4.7) satisfying the additional conditions (2.4.8) and (2.4.9) then h has the form given in (2.4.4) and conversely. Prove this result.

2.15 (Campbell, 1974). Let P and Q be probability measures on a space r and let P_n and Q_n be the corresponding product measures on the product space r^n. For $0 < \lambda < 1$, let $b(n,\lambda) = \inf Q_n(E)$ where the infimum is over all events $E \subset r^n$ for which $P_n(E) \geq \lambda$. Show that, as $n \to \infty$, $b(n,\lambda) = \exp[-n(I(P:Q) + 0(1)]$, where $I(P:Q)$ is the directed divergence.

CHAPTER 3

THE CONCEPT OF INACCURACY

3.0 INTRODUCTION

In this chapter we will discuss the concept of inaccuracy. The idea of entropy is generalized by inaccuracy. Further generalizations of inaccuracy are also discussed. Some applications of the concept of inaccuracy, its continuous analogues and open problems are also discussed.

The characterization theorems of this chapter are taken from Rathie and Kannappan (1971a,1973) and Kaufman and Rathie (1974). The exercises at the end of the chapter are taken from Kannappan (1972,1972a), Kannappan and Rathie (1971), Kerridge (1961) and Rathie (1974).

3.1 INACCURACY AND ITS AXIOMATIC CHARACTERIZATION

Consider two finite discrete probability distributions $P=(p_1,\ldots,p_n)$, $p_i \geq 0$, $\Sigma p_i =1$ and $Q=(q_1,\ldots,q_n)$, $q_i \geq 0$, $\Sigma q_i =1$, i.e., $P \in S_n$ and $Q \in S_n$. Suppose that an observer assigns the distribution Q to an experiment with the sample space partitioned into n mutually exclusive events whereas the true underlying distribution is P, then it is proposed by Kerridge (1961) that the inaccuracy of the statement of the observer can be measured by the following:

(a) DEFINITION 3.1.1

(3.1.1) $H_n(P:Q) = - \Sigma_{i=1}^{n} p_i \, log_2 \, q_i$

Here we will follow the convention that whenever a $q_i=0$ the corresponding $p_i = 0$ and that $0 \, log \, 0 = 0$. For $n=2$, (3.1.1) reduces to

(3.1.2) $H_2(p,1-p:q,1-q) = - p \, log \, q - (1-p) \, log \, (1-q)$

where $(p,q) \in J =]0,1[\times]0,1[\cup\{(0,y)\}\cup\{(1,y')\}$, with $y \in [0,1[$ and $y' \in]0,1]$. We will call (3.1.2) the inaccuracy function.

(b)   PROPERTIES

As the measure of inaccuracy is a generalization of Shannon's entropy, several properties will be common for the two quantities. The various properties for inaccuracy are listed below without proof.

(i)   Non-negativity: For $P \in S_n$, $Q \in S_n$,

(3.1.3) $H_n(P:Q) \geq 0$

with equality if and only if $p_i = q_i = 1$ for one value and $p_i = q_i = 0$ for all other i.

(ii)   Expansibility: For $P \in S_n$, $Q \in S_n$,

(3.1.4) $H_{n+1}(p_1,\ldots,p_n,0:q_1,\ldots,q_n,0) = H_n(P:Q)$

(iii)   Symmetry: For $P \in S_n$, $Q \in S_n$,

(3.1.5) $H_n(p_{a_1},\ldots,p_{a_n}:q_{a_1},\ldots,q_{a_n}) = H_n(p_1,\ldots,p_n:q_1,\ldots,q_n)$

where $\{a_1,\ldots,a_n\}$ is an arbitrary permutation of $\{1,\ldots,n\}$.

(iv)   Recursivity: For $P \in S_n$, $Q \in S_n$ for all $n = 3,4,\ldots$ and $p_1+p_2 > 0$, $q_1+q_2 > 0$,

(3.1.6) $H_n(p_1,\ldots,p_n:q_1,\ldots,q_n) = H_{n-1}(p_1+p_2,p_3,\ldots,p_n:q_1+q_2,q_3,\ldots,q_n)$

$\qquad + (p_1+p_2) H_2(p_1/(p_1+p_2),p_2/(p_1+p_2):q_1/(q_1+q_2),q_2/(q_1+q_2))$

(v)   Additivity: For $p_i, q_j, r_i, s_j \geq 0$, $i = 1,\ldots,m$, $j = 1,\ldots,n$, $\Sigma_{i=1}^{m} p_i = 1$, $\Sigma_{j=1}^{n} q_j = 1$, $\Sigma_{i=1}^{m} r_i = 1$, $\Sigma_{j=1}^{n} s_j = 1$,

(3.1.7) $H_{mn}(p_1 q_1,\ldots,p_1 q_n,\ldots,p_m q_1,\ldots,p_m q_n : r_1 s_1,\ldots,r_1 s_n,\ldots,r_m s_1,\ldots,r_m s_n)$

$\qquad = H_m(p_1,\ldots,p_m:r_1,\ldots,r_m) + H_n(q_1,\ldots,q_n:s_1,\ldots,s_n)$

(vi)   Strong additivity: For $\Sigma_{i=1}^{n} \Sigma_{j=1}^{m} p_{ij} = 1$, $p_j = \Sigma_{i=1}^{n} p_{ij} > 0$, $q_j = \Sigma_{i=1}^{n} q_{ij} > 0$, $\Sigma_{i=1}^{n} \Sigma_{j=1}^{m} q_{ij} = 1$,

(3.1.8) $H_{mn}(p_{11},\ldots,p_{nm}:q_{11},\ldots,q_{nm}) = H_m(p_1,\ldots,p_m:q_1,\ldots,q_m)$

$\qquad + \Sigma_{j=1}^{m} p_j H_n(p_{1j}/p_j,\ldots,p_{nj}/p_j : q_{1j}/q_j,\ldots,q_{nj}/q_j)$

(vii)   Continuity: $H_n(p_1,\ldots,p_n:q_1,\ldots,q_n)$ is a continuous function of its $2n$ variables.

Axiomatic Characterization

(viii) Functional Equation: Denoting $H_2(p,1-p:q,1-q)$ by $f(p,q)$, it is easy to see that f satisfies the functional equation,

$$f(x,y) + (1-x)f(u/(1-x),v/(1-y)) = f(u,v) + (1-u)f(x/(1-u),y/(1-v))$$

for $x,y,u,v \in [0,1[$ with $x+u, y+v \in [0,1]$.

(c) CHARACTERIZATION THEOREM

In this section we will give an axiomatic characterization of the inaccuracy *(3.1.1)* due to Rathie and Kannappan (1971a). Consider the following postulates on H_n.

A1. *Recursivity:* (3.1.6).

A2. *Symmetry:* (3.1.5) for $n = 3$.

A3. *Derivative:* Let $f(p,q) = H_2(p,1-p:q,1-q)$ for $(p,q) \in J$ have continuous first partial derivatives with respect to both the variables $p,q \in]0,1[$ with $f(½,½) = 1$.

A4. *Normalization:* $H_2(1/4,3/4:1/2,1/2) = 1$.

THEOREM 3.1.1: The function $H_n(P:Q)$ for $P \in S_n$, $Q \in S_n$ satisfying the postulates $A1, A2, A3$ and $A4$ is the inaccuracy given by *(3.1.1)* with the usual convention.

Proof: As before, we can prove results equivalent to lemmas 2.1.1, 2.1.2 and 2.1.3 for $H_n(P:Q)$. Here we give a few such results which will be useful in the proof of theorem 3.1.1. Since the derivations are similar to the corresponding results in chapter 2 the details are omitted.

(3.1.9) $\quad H_2(p_1/(p_1+p_2),p_2/(p_1+p_2):q_1/(q_1+q_2),q_2/(q_1+q_2))$

$\quad\quad\quad = H_2(p_2/(p_1+p_2),p_1/(p_1+p_2):q_2/(q_1+q_2),q_1/(q_1+q_2))$

for $p_1+p_2 > 0$, $q_1+q_2 > 0$.

(3.1.10) $\quad f(0,0) = f(1,1)$

where $f(p,q)$ is defined in postulate $A3$.

(3.1.11) $\quad f(p_1+p_2,q_1+q_2) + (p_1+p_2)f(p_1/(p_1+p_2),q_1/(q_1+q_2))$

$\quad\quad\quad = f(p_1,q_1) + (1-p_1)f(p_2/(1-p_1),q_2/(1-q_1))$

$\quad\quad\quad = f(p_2,q_2) + (1-p_2)f(p_1/(1-p_2),q_1/(1-q_2))$

for $p_1,p_2,q_1,q_2 \in [0,1[$, $p_1+p_2, q_1+q_2 \in]0,1]$.

(3.1.12) $\quad f(x,y) = a[x \log(x/2) + (1-x) \log\{(1-x)/2\}] + bx \log\{y/(1-y)\} + g(y)$

for $x,y \in \,]0,1[\,$, where g is a function of y alone. Substituting $x=\frac{1}{4}$, $y=\frac{1}{2}$ in (3.1.12) and utilizing postulate A4, we get

(3.1.13) $a[-3/4 + (3/4)\log 3] + g(1/2) = 1$

The equation (3.1.12) with $x=\frac{1}{2}$, $y=\frac{1}{2}$ and $f(\frac{1}{2},\frac{1}{2})=1$ give

(3.1.14) $-2a + g(1/2) = 1$

Thus (3.1.13) and (3.1.14) give $a=0$. Hence

(3.1.15) $f(x,y) = bx \log(y/(1-y)) + g(y)$

The last two equations of (3.1.11) for $p_1=0$, $q_1=0$ give $f(0,0)=0$ and hence from (3.1.10), we have

(3.1.16) $f(0,0) = f(1,1) = 0$

For $p_1=q_1=p$ and $p_2=q_2=q$ the second and the third equation-pair in (3.1.11) give

(3.1.17) $G(p) + (1-p)G(q/(1-p)) = G(q) + (1-q)G(p/(1-q))$

for $p,q \in [0,1[\,$, $p+q \in \,]0,1]$, where $G(p) = f(q,p)$. Now when $p+q=0$ then both p and q are zeros and (3.1.17) is satisfied. Hence equation (3.1.17) is valid for $p,q \in [0,1[\,$, $p+q \in [0,1]$. Also by postulate A3, f is differentiable and therefore f is continuous and so is G in $\,]0,1[\,$. Thus (3.1.17), (3.1.16) and $f(\frac{1}{2},\frac{1}{2})=1$ give (see Daróczy, 1969a),

(3.1.18) $f(x,x) = -x \log x - (1-x)\log(1-x)$, $x \in [0,1]$

Hence (3.1.15) with $y=x$ and (3.1.18) give

(3.1.19) $g(x) = -x \log x - (1-x)\log(1-x) - bx \log\{x/(1-x)\}$

Thus (3.1.15) and (3.1.19) yield

(3.1.20) $f(x,y) = (bx-y-by) \log\{y/(1-y)\} - \log(1-y)$, $x,y \in \,]0,1[$

For $p_1 = q_1 = 1/2$, $p_2 = 1/8$, $q_2 = 1/4$, the second and the third equation pair in (3.1.11) yield

(3.1.21) $f(1/2,1/2) + (1/2)f(1/4,1/2) = f(1/8,1/4) + (7/8)f(4/7,2/3)$

Now from $f(1/2,1/2) = 1$, postulate A4, (3.1.20) and (3.1.21) we have $b=-1$ and therefore (3.1.20) becomes

(3.1.22) $f(x,y) = -x \log y - (1-x) \log(1-y)$, $x,y \in \,]0,1[$

Next, we find $f(0,y)$ and $f(1,y)$ for $y \in \,]0,1[$. For $p_1=0$, $q_1=\frac{1}{2}$, $q_2=\frac{1}{4}$, the second and third equation-pair in (3.1.11) yield

$(3.1.23)$ $f(0,1/2) + f(p_2,1/2) = f(p_2,1/4) + (1-p_2)f(0,2/3)$

for $p_2 \in]0,1[$. The equation $(3.1.23)$ with $p_2 = 1/2$ gives

$(3.1.24)$ $f(0,1/2) + f(1/2,1/2) = f(1/2,1/4) + (1/2)f(0,2/3)$

Hence $(3.1.23),(3.1.24),(3.1.22)$ and $f(1/2,1/2) = 1$, give

$(3.1.25)$ $f(0,1/2) = 1$

and

$(3.1.26)$ $f(0,2/3) = \log 3$

The second and third equation-pair in $(3.1.11)$ for $p_1 = 0$ give

$(3.1.27)$ $f(0,q_1) + f(p_2, q_2/(1-q_1)) = f(p_2,q_2) + (1-p_2)f(0, q_1/(1-q_2))$

for $p_2, q_2 \in]0,1[$, $q_1 \in [0,1[$, $q_1 + q_2 \in]0,1]$. The equations $(3.1.27)$ with $q_2 = 1 - 2q_1$, $(3.1.22)$ and $(3.1.25)$ give

$(3.1.28)$ $f(0,q_1) = -\log(1-q_1)$, $q_1 \in]0,\tfrac{1}{2}[$

Again, the equations $(3.1.27)$ with $q_2 = 1/2$, $(3.1.22)$ and $(3.1.28)$ yield

$(3.1.29)$ $f(0,2q_1) = -\log(1-2q_1)$, $2q_1 \in]0,1[$

Hence $(3.1.28)$ and $(3.1.29)$ give

$(3.1.30)$ $f(0,q) = -\log(1-q)$, $q \in]0,1[$

Rewriting $(3.1.9)$, we get

$(3.1.31)$ $f(p,q) = f(1-p, 1-q)$, $(p,q) \in J$

Taking $p=0$ in $(3.1.31)$ and utilizing $(3.1.30)$, we have

$(3.1.32)$ $f(1,q) = -\log q$, $q \in]0,1[$

Hence from $(3.1.22),(3.1.30),(3.1.32)$ and $(3.1.16)$, we get

$(3.1.33)$ $f(x,y) = -x \log y - (1-x) \log(1-y)$, $(x,y) \in J$

Now by induction, postulate A1 implies that

$(3.1.34)$ $H_n(P:Q) = \sum_{i=2}^n r_i\, H_2(r_{i-1}/r_i,\, p_i/r_i : s_{i-1}/s_i,\, q_i/s_i)$

where $r_i = p_1 + \ldots + p_i$ and $s_i = q_1 + \ldots + q_i$, for $i = 1,\ldots,n$. The equations $(3.1.33)$ and $(3.1.34)$ give

$$H_n(P:Q) = -\sum_{i=2}^n r_i\{(p_i/r_i)\log(q_i/s_i) + (r_{i-1}/r_i)\log(s_{i-1}/s_i)\}$$

$$= -\sum_{i=2}^n p_i \log q_i + \sum_{i=2}^n \{r_i \log s_i - r_{i-1} \log s_{i-1}\}$$

$$= -\sum_{i=1}^n p_i \log q_i$$

This completes the proof of Theorem $3.1.1$.

3.2 INACCURACIES OF ORDER α

In this section we introduce the concept of non-additive inaccuracy function of order $\alpha\,(\alpha\neq 1)$ and the additive and non-additive inaccuracies of order α. These concepts generalize the corresponding quantities of chapter 2.

(a) DEFINITIONS

Let $P\epsilon S_n$ and $Q\epsilon S_n$. Then, we define the non-additive inaccuracy of order $\alpha\,(\alpha\neq 1)$ by the following.

Definition 3.2.1

(3.2.1) $\qquad H_{n,\alpha}(P:Q) = (\Sigma\, p_i\, q_i^{\alpha-1} -1)/(2^{1-\alpha} -1),\ \alpha \neq 1$

where the convention $0^{\alpha}=0$, $(\alpha\neq 0)$ is used. For $n=2$, (3.2.1) takes the following form.

(3.2.2) $\qquad H_{2,\alpha}(p,1-p:q,1-q) = \{pq^{\alpha-1} + (1-p)(1-q)^{\alpha-1} -1\}/(2^{1-\alpha} -1),\ \alpha \neq 1$

We will call $H_{2,\alpha}$ as the *non-additive inaccuracy function of order* α $(\alpha\neq 1)$. The additive inaccuracy of order $\alpha(\alpha\neq 1)$ is defined by the following.

Definition 3.2.2

(3.2.3) $\qquad \hat{H}_{n,\alpha}(P:Q) = (1-\alpha)^{-1}\, log(\Sigma p_i q_i^{\alpha-1}),\ \alpha \neq 1$

Both $H_{n,\alpha}$ and $\hat{H}_{n,\alpha}$ are generalizations of H_n given in (3.1.1) and are connected by the relation

(3.2.4) $\qquad H_{n,\alpha}(P:Q) = (2^{(1-\alpha)\hat{H}_{n,\alpha}} -1)/(2^{1-\alpha} -1),\ \alpha \neq 1$

Alternately, we may define the non-additive inaccuracy function of order α $(\alpha\neq 1)$ and the non-additive inaccuracy as follows. Let

(3.2.5) $\qquad f_{\alpha}(p,q) = H_{2,\alpha}(p,1-p:q,1-q)\ ,\ p,q\ \epsilon\ [0,1]$

Definition 3.2.3: A function $f_{\alpha}:[0,1]\times[0,1]\to \mathcal{R}$ is called an inaccuracy function of order $\alpha(\alpha\neq 1)$ if it satisfies the functional equation

(3.2.6) $\qquad f_{\alpha}(x,y) + (1-x)(1-y)^{\alpha-1}\, f_{\alpha}(u/(1-x),v/(1-y))$

$\qquad\qquad = f_{\alpha}(u,v) + (1-u)(1-v)^{\alpha-1}\, f_{\alpha}(x/(1-u),y/(1-v))$

for all $x,y,u,v\ \epsilon\ [0,1[\ ,\ x+u,\ y+v\ \epsilon\ [0,1]$ and the boundary conditions

(3.2.7) $\qquad f_{\alpha}(0,0) = f_{\alpha}(1,1)$

and

(3.2.8) $\qquad f_{\alpha}(\tfrac{1}{2},\tfrac{1}{2}) = 1$

Inaccuracies of Order α   81

Definition 3.2.4: If f_α is an inaccuracy function of order α as given in Definition 2.2.3, then the function $H_{n,\alpha}$, the inaccuracy of order α ($\alpha \neq 1$) is defined by the expression

(3.2.9) $H_{n,\alpha}(p_1,\ldots,p_n:q_1,\ldots,q_n) = \sum_{i=2}^{n} r_i s_i^{\alpha-1} f_\alpha(p_i/r_i, q_i/s_i)$

where $r_i = p_1 + \ldots + p_i$ and $s_i = q_1 + \ldots + q_i$ for $i=1,\ldots,n$, with $r_n = s_n = 1$.

(b) PROPERTIES

(i) Non-negativity: Both $H_{n,\alpha}(P:Q)$ and $\hat{H}_{n,\alpha}(P:Q)$ are non-negative. The equality holds if and only if p_i and q_i are unity for one i and zeros for the rest of the i's.

(ii) Expansibility: $H_{n,\alpha}(p_1,\ldots,p_n,0:q_1,\ldots,q_n,0) = H_{n,\alpha}(p_1,\ldots,p_n:q_1,\ldots,q_n)$. A similar property holds for $\hat{H}_{n,\alpha}$

(iii) Symmetry: $H_{n,\alpha}$ and $\hat{H}_{n,\alpha}$ remain unchanged if the elements of P and Q are rearranged in the same way so that the one-to-one correspondence between them is not changed.

(iv) Recursivity: For $P \in S_n$, $Q \in S_n$ and $p_1+p_2 > 0$, $q_1+q_2 > 0$,

(3.2.10) $H_{n,\alpha}(p_1,\ldots,p_n:q_1,\ldots,q_n) = H_{n-1,\alpha}(p_1+p_2,p_3,\ldots,p_n:q_1+q_2,q_3,\ldots,q_n)$
$+ (p_1+p_2)(q_1+q_2)^{\alpha-1} H_{2,\alpha}(p_1/(p_1+p_2),p_2/(p_1+p_2):q_1/(q_1+q_2),q_2/(q_1+q_2))$

(v) Additivity: For $\sum_{i=1}^{n} p_i = \sum_{i=1}^{n} q_i = \sum_{j=1}^{m} P_j = \sum_{j=1}^{m} Q_j = 1$,

(3.2.11) $\hat{H}_{mn,\alpha}(p_1 P_1,\ldots,p_1 P_m,\ldots,p_n P_1,\ldots,p_n P_m:q_1 Q_1,\ldots,q_1 Q_m,\ldots,q_n Q_1,\ldots$
$\ldots,q_n Q_m) = \hat{H}_{n,\alpha}(p_1,\ldots,p_n:q_1,\ldots,q_n) + \hat{H}_{m,\alpha}(P_1,\ldots,P_m:Q_1,\ldots,Q_m)$

(vi) Non-additivity: For $\sum_{i=1}^{n} p_i = \sum_{i=1}^{n} q_i = \sum_{j=1}^{m} P_j = \sum_{j=1}^{m} Q_j = 1$,

(3.2.12) $H_{mn,\alpha}(p_1 P_1,\ldots,p_1 P_m,\ldots,p_n P_1,\ldots,p_n P_m:q_1 Q_1,\ldots,q_1 Q_m,\ldots,q_n Q_1,\ldots$
$\ldots,q_n Q_m) = H_{n,\alpha}(p_1,\ldots,p_n:q_1,\ldots,q_n) + H_{m,\alpha}(P_1,\ldots,P_m:Q_1,\ldots,Q_m)$
$+ (2^{1-\alpha} - 1) H_{n,\alpha}(p_1,\ldots,p_n:q_1,\ldots,q_n) H_{m,\alpha}(P_1,\ldots,P_m:Q_1,\ldots,Q_m)$

(vii) Strong Non-additivity: For $\sum_{i=1}^{n} p_i = \sum_{i=1}^{n} q_i = 1$, $\sum_{j=1}^{m} P_{ji} = 1$ and $\sum_{j=1}^{m} q_{ji} = 1$ for all $i = 1,\ldots,n$,

(3.2.13) $H_{mn,\alpha}(p_1 p_{11},\ldots,p_1 p_{m1},\ldots,p_n p_{1n},\ldots,p_n p_{mn}:q_1 q_{11},\ldots,q_1 q_{m1},\ldots$
$\ldots,q_n q_{1n},\ldots,q_n q_{mn}) = H_{n,\alpha}(p_1,\ldots,p_n:q_1,\ldots,q_n)$

$$+ \Sigma_{i=1}^{n} p_i q_i^{\alpha-1} H_{m,\alpha}(p_{1i},\ldots,p_{mi}:q_{1i},\ldots,q_{mi})$$

The property *(vii)* is a generalization of the property *(vi)*.

(viii) Continuity: $H_{n,\alpha}$ and $\hat{H}_{n,\alpha}$ are continuous functions of their variables.

(c) CHARACTERIZATION THEOREMS

In this section we will state without proof two characterization theorems due to Rathie and Kannappan (1973) regarding the non-additive inaccuracy function and inaccuracy of order α defined in *(3.2.2)* and *(3.2.1)* respectively. Also we will discuss in detail a theorem due to Kaufman and Rathie (1974).

THEOREM 3.2.1: If $f_\alpha(x,y)$ is an inaccuracy function of order $\alpha(\alpha \neq 1)$ as given in definition 3.2.3 then $f_\alpha(x,y)$ is given by *(3.2.2)* for all $x,y \in [0,1]$ and $\alpha \neq 1$.

THEOREM 3.2.2: If $H_{n,\alpha}(P:Q)$, for $n = 2,3,\ldots,$ satisfies
B1. *Recursivity:* (3.2.10),
B2. *Symmetry:* property 3.2 *(iii)* for $n = 3$,
B3. *Normalization:* $H_{2,\alpha}(\frac{1}{2},\frac{1}{2}:\frac{1}{2},\frac{1}{2}) = 1$,
then $H_{n,\alpha}$ is uniquely determined and is given by *(3.2.1)* for $\alpha \neq 1$.

The following theorem is taken from Kaufman and Rathie (1974).

THEOREM 3.2.3: The general measurable solutions of *(3.2.6)* are given by

(3.2.14) $\quad f_\alpha(x,y) = a\{(1-x)(1-y)^\alpha - 1\} + b\ xy^\alpha$

for $x,y \in [0,1]$, where a and b are arbitrary constants and $\alpha(>0) \neq 1$.

Proof: The functional equation *(3.2.6)* for each specified $y, v \in [0,1[$ with $y+v \in [0,1]$ takes the following form.

(3.2.15) $\quad F(x) + (1-x)G(y/(1-x)) = H(y) + (1-y)K(x/(1-y))$

for all $x, y \in [0,1[$ with $x+y \in [0,1]$. Hence, using the results due to Kannappan and Ng (1973), we have

(3.2.16) $\quad f_\alpha(x,y) = A(y,v)f(x) + B_1(y,v)x + D(y,v), \quad x \in [0,1[$

(3.2.17) $\quad (1-y)^\alpha f_\alpha(x,v/(1-y)) = A(y,v)f(x) + B_2(y,v)x + B_1(y,v) - B_4(y,v),$
$$x \in [0,1]$$

(3.2.18) $\quad f_\alpha(x,v) = A(y,v)f(x) + B_3(y,v)x + B_1(y,v) + B_2(y,v) - B_3(y,v)$
$\quad\quad\quad - B_4(y,v) + D(y,v), \quad x \in [0,1[$

(3.2.19) $(1-v)^{\alpha}f_{\alpha}(x,y/(1-v)) = A(y,v)f(x) + B_4(y,v)x + B_3(y,v) - B_2(y,v)$,

for $x \in [0,1]$ where $f(x)$ is given by (1.1.3) and the constants A, B_1, B_2, B_3, B_4 and D are expressed as functions of y and v.

Following an argument similar to that followed in chapter 2, it is easy to see that

(3.2.20) $f_{\alpha}(x,y) = \hat{A} S(x) + \hat{B}(y)x + \hat{D}(y)$, $x,y \in [0,1]$

where $A(y,v) = \hat{A}$, a constant, $B_1(y,v) = \hat{B}(y)$ and $D(y,v) = \hat{D}(y)$. After some simplifications one gets

(3.2.21) $\hat{A} = 0$

(3.2.22) $(1-y)^{\alpha}\hat{D}(v/(1-y)) = \hat{B}(y) - (1-v)^{\alpha}\hat{B}(y/(1-v))$

and
(3.2.23) $\hat{D}(y) + (1-y)^{\alpha}\hat{D}(v/(1-y)) = \hat{D}(v) + (1-v)^{\alpha}\hat{D}(y/(1-v))$

for $y, v \in [0,1[$ with $y+v \in [0,1]$ and the function $\hat{D}$ is measurable.

The measurable solutions $\hat{D}(y)$ of (3.2.23) are obtained in chapter 2 as

(3.2.24) $\hat{D}(x) = a f_{\alpha}(x) + cx^{\alpha}$, $x \in [0,1]$, $\alpha(>0) \neq 1$,

where $f_{\alpha}(x)$ is given by (1.2.5). Using this expression for $\hat{D}(y)$ in (3.2.22) and after some simplifications, one can get

(3.2.25) $\hat{B}(y) = -a(1-y)^{\alpha} + b y^{\alpha}$, $y \in]0,1[$

and
(3.2.26) $a + c = 0$

Hence (3.2.20), (3.2.21), (3.2.24) and (3.2.25) give $f_{\alpha}(x,y)$ for $x \in [0,1]$, $y \in]0,1[$, $\alpha(>0) \neq 1$. It is easy to find $f_{\alpha}(x,0)$ and $f_{\alpha}(x,1)$. Thus the theorem 3.2.3 is proved.

3.3 PSEUDO-MEASURES

Let $P \in S_n$ and let $Q = (q_1, \ldots, q_n)$, $q_i \geq 0$, $\Sigma_{i=1}^{n} q_i \leq 1$. Then the pseudo-inaccuracy can be defined by the expression

(3.3.1) $\hat{H}_n(p_1, \ldots, p_n; q_1, \ldots, q_n) = -\Sigma_{i=1}^{n} p_i \log q_i$

with the usual convention $0 \log 0 = 0$.

Kerridge (1961) claims to have obtained a characterization theorem for (3.1.1) but in fact his postulate 4 violates the condition $\Sigma \phi_i = 1$. So his result may be regarded as giving an axiomatic characterization of (3.3.1) instead of (3.1.1). His theorem is put as an exercise at the end of this chapter. For continuous and Lebesgue measurable solutions of certain functional equations concerning pseudo-inaccuracy, see Kannappan (1972,1972a). Similar pseudo-measures corresponding to the generalized measures of inaccuracy can be defined and studied.

3.4 FURTHER GENERALIZATIONS

Some generalizations of *(3.1.1)*, *(3.2.1)* and *(3.2.3)* are available in recent literature. The following generalizations are due to Rathie (1970a).

(3.4.1) $\quad H_1(p_1,\ldots,p_n;\beta_1,\ldots,\beta_n;q_1,\ldots,q_n) = -\Sigma p_i^{\beta_i} \log q_i / \Sigma p_i^{\beta_i}$

(3.4.2) $\quad H_\alpha(p_1,\ldots,p_n;\beta_1,\ldots,\beta_n;q_1,\ldots,q_n) = (\alpha-1)^{-1} \log(\Sigma p_i^{\beta_i} q_i^{\alpha-1} / \Sigma p_i^{\beta_i}), \alpha \neq 1$

Some interesting special cases of *(3.4.1)* and *(3.4.2)* can be obtained by taking $\beta_1 = \beta_2 = \ldots = \beta_n = \beta$. For further details one may go through Rathie (1970a,1971a) and Rathie and Nath (1972).

3.5 CONTINUOUS ANALOGUES

The continuous analogue to inaccuracy *(3.1.1)* for the two probability measures P and Q with density functions $p(x)$ and $q(x)$ respectively is defined by

(3.5.1) $\quad H(P:Q) = - \int p(x) \log q(x) \, dx$

In a similar manner other continuous analogues can be defined and their properties studied.

3.6 APPLICATIONS

The quantity $-\Sigma p_i \log q_i$ can be interpreted as the *equivocation of the inference* in Bayes' inference problems. For example, suppose that θ_i is the probability that the i-th hypothesis is true with $\Sigma \theta_i = 1$. Let α_{ij} be the probability that the j-th result of an experiment is observed if the i-th hypothesis is true. Suppose that the experimenter assumes β_{ij} to be the probability that the i-th hypothesis is true when the j-th outcome is observed then the average inaccuracy of the inference is $-\Sigma_{i,j} \theta_i \alpha_{ij} \log \beta_{ij}$. (see Kerridge 1961).

The quantity $-\Sigma p_i \log q_i$ may also be interpreted as a measure of information in an experiment - see Mallows (1959).

In psychological problems, one can give the following inter-

Applications

pretation to $\Sigma p_i q_i$. Suppose that the probability of correct anticipation of a symbol i is q_i and the corresponding true probability is p_i. Then $\Sigma p_i q_i$ is the expected anticipation which may be called a measure of *discriminating power* of psychological tests. This method in Psychology is usually known as *dispersion analysis for categorical data* - see Quastler (1956).

3.7 OPEN PROBLEMS

Derive characterization theorems for *(3.4.1)* and *(3.4.2)* by assuming different sets of postulates for each of these measures.

EXERCISES

3.1 (Kerridge, 1961). The only $H_n(p_1,\ldots,p_n;q_1,\ldots,q_n)$, $p_i \geqslant 0$, $\Sigma p_i = 1$, $q_i \geqslant 0$, $\Sigma q_i \leqslant 1$ satisfying the following postulates is $H_n = -k \Sigma p_i \log q_i$, where k is a positive constant: *(a)* The function H_n is continuous in p_i's and q_i's; *(b)* When n alternatives are stated to be equally likely, the inaccuracy is a monotonic increasing function of n; *(c)* If a statement is broken down into a number of subsidiary statements, the inaccuracy of the original statement is a weighted sum of the inaccuracies of the subsidiary statements; *(d)* The inaccuracy of a statement is unchanged if two alternatives about which the same assertion is made are combined.

3.2 (Kannappan and Rathie 1971). Solve the functional equation $f(x,y) + g(x) h(y) f(u/(1-x), v/(1-y)) = f(u,v) + g(u) h(v) f(x/(1-u), y/(1-v))$, for $x,y,u,v \in [0,1[$, $x+u$, $y+v \in I=[0,1]$ and $g:I \to \mathcal{R}$ (reals) and $h:I \to \mathcal{R}$ are given functions satisfying the functional equation $\phi(x+y-xy) = \phi(x)\phi(y)$, for $x,y \in I$, and $f:I \times I \to \mathcal{R}$, satisfying further $f(0,0) = f(1,1)$ and $f(\tfrac{1}{2},\tfrac{1}{2}) = 1$.

3.3 Prove theorems *3.2.1* and *3.2.2*.

3.4 (Rathie, 1974). Find the measurable solutions of the functional equation $F(x,y) + (1-x)(1-y)F(u/(1-x), v/(1-y)) = F(u,v) + (1-u)(1-v) F(x/(1-u), y/(1-v))$, for $x,y,u,v \in [0,1[$ with $x+u$, $y+v \in [0,1]$.

3.5 (Kannappan, 1972a). Show that the most general Lebesgue measurable solution F defined on J of the functional equation $\Sigma_{i=1}^{2}\Sigma_{j=1}^{3} F(x_i y_j, u_i v_j) = \Sigma_{i=1}^{2} F(x_i, u_i) + \Sigma_{j=1}^{3} F(y_j, v_j)$, for $x_i, y_j, u_i, v_j \geqslant 0$, $\Sigma_{i=1}^{2} x_i = 1$, $\Sigma_{j=1}^{3} y_j = 1$, $\Sigma_{i=1}^{2} u_i \leqslant 1$, $\Sigma_{j=1}^{3} v_j \leqslant 1$, satisfying the conditions $F(\tfrac{1}{2},\tfrac{1}{2}) = \tfrac{1}{2}$, $F(1,\tfrac{1}{2}) = 1$ and $F(1,1) = 0$ has the form $F(x,y) = -x \log y$.

3.6 (Kannappan, 1972). Prove that $F(x,y) = -x \log y$, $(x,y) \in J$ is a continuous solution of the functional equation,

$\sum_{i=1}^{m} \sum_{j=1}^{n} F(x_i y_j, u_i v_j) = \sum_{i=1}^{m} F(x_i, u_i) + \sum_{j=1}^{n} F(y_j, v_j)$, for $x_i, u_i, y_j, v_j \geq 0$, $\sum_{i=1}^{m} x_i = 1$, $\sum_{j=1}^{n} y_j = 1$, $\sum_{i=1}^{m} u_i \leq 1$ and $\sum_{j=1}^{n} v_j \leq 1$ under the conditions $F(\frac{1}{2}, \frac{1}{2}) = \frac{1}{2}$ and $F(1, \frac{1}{2}) = 1$.

CHAPTER 4

SOME BASIC STATISTICAL CONCEPTS AND THEIR CHARACTERIZATIONS

4.0 INTRODUCTION

This chapter deals mainly with the following basic statistical concepts: (1) Covariance between two stochastic variables; (2) The concept of affinity between two statistical distributions; (3) Pearson's chi-square measure of discrepancy between two discrete distributions; (4) A general theory of dispersion or scatter in a statistical population; (5) Distance between two statistical populations. In statistical literature, these as well as other related basic concepts are defined mainly because of their practical uses in analysing data or in making a decision by using the available information contained in a given data. At present, these definitions have only intuitive justifications and statisticians put more emphasis on the practical utility of these concepts rather than their mathematical foundations.

In this chapter some of these measures will be discussed in detail. A number of postulates are given for these measures, which are in fact properties of corresponding measures used by statisticians and thus these are desirable properties from a statistical point of view. From a mathematical point of view these properties can act as a set of axioms which will characterize the various concepts. Thus through these axiomatic characterizations some of the basic concepts will be put on mathematical foundations. Characterizations are given for the concepts of covariance, affinity, some measures of distance between two populations and Pearson's measure of discrepancy. These can also act as axiomatic definitions of these measures. All the theorems given in this chapter are taken from Mathai and Rathie (1971, 1971a) and Kaufman, Mathai and Rathie (1971).

4.1 COVARIANCE AND ITS AXIOMATIC FOUNDATION

Consider a set of n paired observations $(x_1,y_1),\ldots,(x_n,y_n)$ such as height and weight measurements of a sample of n individuals. In order to study the joint variation in $(x_1,\ldots,x_n)$ and $(y_1,\ldots,y_n)$ statisticians often use a measure called covariance between x's and y's. The main aim of this section is to characterize covariance by using a set of axioms. These axioms are also desirable properties from a statistical point of view in the sense that they are satisfied by a measure of joint variation, namely, covariance which is already used by statisticians and which is constructed from heuristic considerations. A characterization of variance will also be given in this section.

(a) *DEFINITION:* If $(x_1,y_1),\ldots,(x_n,y_n)$ is a set of n paired observations then the covariance between x's and y's is defined as

(4.1.1) $\quad Cov(x,y) = \sum_{i=1}^{n}(x_i - \bar{x})(y_i - \bar{y})/n$

where $\bar{x} = (x_1+\ldots+x_n)/n$ and $\bar{y} = (y_1+\ldots+y_n)/n$. If X and Y are two stochastic variables having a joint distribution then the covariance between X and Y is defined as,

(4.1.2) $\quad Cov(X,Y) = E[\{X-E(X)\}\{Y-E(Y)\}]$

where E denotes the operator *mathematical expectation*.

(b) *CHARACTERIZATION:* A characterization of (4.1.1) is given here by using six postulates. Let $f(x_1,\ldots,x_n:y_1,\ldots,y_n)$ be a function of x's and y's satisfying the following postulates. For x_i, y_i, $z_i \in (-\infty,\infty)$, $i = 1,2,\ldots,n$, let

A1. $f(x_1+z_1,\ldots,x_n+z_n:y_1,\ldots,y_n) = f(x_1,\ldots,x_n:y_1,\ldots,y_n) + f(z_1,\ldots,z_n:y_1,\ldots,y_n)$;

A2. $f(x_1,\ldots,x_n:y_1,\ldots,y_n)$ is a continuous function of its variables;

A3. $f(x_1,\ldots,x_n:y_1,\ldots,y_n) = f(y_1,\ldots,y_n:x_1,\ldots,x_n)$;

A4. $f(x_1,\ldots,x_n:y_1,\ldots,y_n)$ is symmetric in pairs (x_i,y_i) and (x_j,y_j) for all i and j;

A5. $f(x,x,\ldots,x:y_1,\ldots,y_n) = 0$ for all x;

A6. $f(0,1,1,\ldots,1:0,1,\ldots,1) = (n-1)n^{-2}$.

THEOREM 4.1.1: The postulates A1 to A6 uniquely determine $f(x_1,\ldots,x_n:y_1,\ldots,y_n)$ as the covariance given in (4.1.1).

Proof: It is easy to see from postulates A1 and A2 and Aczél(1966), p.215 that

Covariance and Its Axiomatic Foundation

(4.1.3) $\quad f(x_1,\ldots,x_n:y_1,\ldots,y_n) = x_1 g_1(y_1,\ldots,y_n)+\ldots+ x_n g_n(y_1,\ldots,y_n)$

where $g_1,\ldots,g_n$ are arbitrary functions of $y_1,\ldots,y_n$. Clearly (4.1.3) gives

(4.1.4) $\quad \begin{cases} f(x_1,0,\ldots,0:y_1,\ldots,y_n) = x_1 g_1(y_1,\ldots,y_n) \\ \cdots\cdots\cdots\cdots\cdots\cdots\cdots\cdots\cdots\cdots\cdots\cdots\cdots \\ f(0,\ldots,0,x_n:y_1,\ldots,y_n) = x_n g_n(y_1,\ldots,y_n) \end{cases}$

For $x_1=\ldots=x_n=1$ in (4.1.4) and A4 give

(4.1.5) $\quad g_1(y_1,\ldots,y_n) = g_2(y_2,y_1,\ldots,y_n) = \ldots = g_n(y_n,y_2,\ldots,y_{n-1},y_1)$

In (4.1.3) taking $x_1=\ldots=x_n$ and using A5 one gets

(4.1.6) $\quad g_1(y_1,\ldots,y_n) + g_2(y_1,\ldots,y_n) + \ldots + g_n(y_1,\ldots,y_n) = 0$

Also (4.1.3) and A3 yield

(4.1.7) $\quad f(x_1,\ldots,x_n:y_1,\ldots,y_n) = y_1 g_1(x_1,\ldots,x_n)+\ldots+y_n g_n(x_1,\ldots,x_n)$

From (4.1.3) and (4.1.7) we have

(4.1.8) $\quad x_1 g_1(y_1,\ldots,y_n)+\ldots+x_n g_n(y_1,\ldots,y_n) = y_1 g_1(x_1,\ldots,x_n) + \ldots$
$$\ldots + y_n g_n(x_1,\ldots,x_n)$$

Now by putting $x_1 = x_2 = \ldots = x_{n-1}$ in (4.1.8), subtracting x_1 times (4.1.6) and $\Sigma_{i=1}^{n-1} y_i/(n-1)$ times (4.1.6) with $y_1=\ldots= y_{n-1} = x_1$ and $y_n=x_n$ from the left hand side and right hand side respectively of (4.1.8) and using (4.1.5) one gets

(4.1.9) $\quad (x_n - x_1)g_n(y_1,\ldots,y_n) = (y_n - \bar{y})[n/(n-1)]g_n(x_1,\ldots,x_1,x_n)$

giving

(4.1.10) $\quad g_n(y_1,\ldots,y_n) = c[n/(n-1)](y_n - \bar{y})$

where c is an arbitrary constant. From (4.1.5), $g_1,\ldots,g_{n-1}$ are also obtained. Now subtracting $\bar{x}$ times (4.1.6) from the right hand side of (4.1.3) and using (4.1.10) and (4.1.5) we get

(4.1.11) $\quad f(x_1,\ldots,x_n:y_1,\ldots,y_n) = c\, n(n-1)^{-1} \Sigma_{i=1}^{n} (x_i - \bar{x})(y_i - \bar{y})$

From postulate A6 we get $c = (n-1)n^{-2}$ and thus the proof of theorem 4.1.1 is complete.

The above theorem can also be proved by using different sets of postulates. Also (4.1.2) can be characterized once the form is assumed. For these and other related aspects see the problems at the end of this chapter.

Another important measure which is used by statisticians is the measure of variance. If $x=(x_1,\ldots,x_n)$ is a sample of n observations then the variance of x is defined as

$$(4.1.12) \quad Var(x) = \Sigma(x_i - \bar{x})^2/n$$

A characterization of (4.1.12) can be given by using the following postulates. Let $V(x_1,\ldots,x_n)$ be a function of $x_1,\ldots,x_n$ such that

B1. $V(x_1+y_1,\ldots,x_n+y_n) = V(x_1,\ldots,x_n) + V(y_1,\ldots,y_n) + 2g(x_1,\ldots,x_n;y_1,\ldots,y_n)$
for $x_i, y_i \in (-\infty, \infty)$ for all i where g satisfies the postulates A1 to A6 of theorem 4.1.1;

B2. $V(x_1,\ldots,x_n)$ is symmetric in its variables;

B3. $V(x,\ldots,x) = 0;$

B4. $V(x_1,\ldots,x_n)$ is continuous.

THEOREM 4.1.2: A non-negative function $V(x_1,\ldots,x_n)$ satisfying postulates B1 to B4 uniquely determines the variance given in (4.1.12).

Proof: From theorem 4.1.1, g is given by

$$(4.1.13) \quad g(x_1,\ldots,x_n;y_1,\ldots,y_n) = \Sigma(x_i - \bar{x})(y_i - \bar{y})/n$$

Taking

$$(4.1.14) \quad F(x_1,\ldots,x_n) = V(x_1,\ldots,x_n) - \Sigma(x_i - \bar{x})^2/n$$

in postulate B1 one gets the generalized Cauchy functional equation

$$(4.1.15) \quad F(x_1+y_1,\ldots,x_n+y_n) = F(x_1,\ldots,x_n) + F(y_1,\ldots,y_n)$$

of which the only continuous solution is,

$$(4.1.16) \quad F(x_1,\ldots,x_n) = c_1 x_1 + \ldots + c_n x_n$$

From B2 and B3, $c_1=\ldots=c_n=0$. This completes the proof. For other sets of postulates which will characterize the variance and for a characterization of linear correlation coefficient, see the problems at the end of this chapter.

4.2 THE CONCEPT OF AFFINITY AND ITS CHARACTERIZATIONS

The concept of affinity between two statistical populations was introduced and a number of its properties and applications in statistical decision making were studied by Matusita(1954,1955, 1957,1961,1967) and Kirmani(1968). In this section two different characterizations will be given for the measure of affinity between two discrete distributions. One theorem is based on recursivity principle and the other is based on a maximization principle. It will be too lengthy to point out the various applications of this

The Concept of Affinity and Its Characterizations

concept of affinity. Most of the applications may be seen from the work of Matusita, some of which are listed above.

(a) DEFINITION: Consider two discrete distributions $(p_1,\ldots,p_k)$ and $(q_1,\ldots,q_k)$, $p_i \geq 0$, $q_i \geq 0$, $i = 1,2,\ldots,k$, $\Sigma p_i = \Sigma q_i = 1$. Then

(4.2.1) $\quad \rho(p_1,\ldots,p_k:q_1,\ldots,q_k) = \Sigma_{i=1}^{k}(p_i q_i)^{1/2}$

is defined as the affinity between the two distributions. This can also be interpreted as $\cos\theta$, where θ is the angle between the vectors if the two distributions are represented as points on a circle. In this sense ρ measures the association between the distributions. In the continuous case, if $p(x)$ and $q(x)$ are the density functions of two stochastic variables X and Y, then the affinity between $p(x)$ and $q(x)$ is defined as,

(4.2.2) $\quad \rho = \int_{-\infty}^{\infty}[p(x)q(x)]^{1/2}dx$

This measure is also called Bhattacharyya coefficient (Bhattacharyya, 1945-46). It is also generalized to define the affinity among a number of distributions. Some problems are cited in this direction at the end of this chapter.

(b) CHARACTERIZATIONS: The first theorem is based on recursivity, symmetry and normalization postulates. These postulates are also intuitively justifiable. Let $\rho_n(p_1,\ldots,p_n:q_1,\ldots,q_n)$ be a function of $p_1,\ldots,p_n$ and $q_1,\ldots,q_n$ where p_i, $q_i \geq 0$, $i = 1,2,\ldots,n$, $\Sigma p_i = 1 = \Sigma q_i$ and satisfying the following postulates.

C1. *Recursivity:* $\rho_n(p_1,\ldots,p_n:q_1,\ldots,q_n) = \rho_{n-1}(p_1+p_2,p_3,\ldots,p_n:q_1+q_2,q_3,\ldots,q_n)$
$+ (p_1+p_2)^{1/2}(q_1+q_2)^{1/2}[\rho_2(p_1/(p_1+p_2),p_2/(p_1+p_2):q_1/(q_1+q_2),q_2/(q_1+q_2))-1]$
for all $n > 2$, p_1+p_2, $q_1+q_2 > 0$;

C2. *Symmetry:* ρ_3 is symmetric in pairs $\{p_i,q_i\}$, $i = 1,2,3$;

C3. *Normalization:* $\rho_2(1/4,3/4:3/4,1/4) = \cos(\pi/6)$.

Postulate *C1* explains the desired nature of combinations of the measures to be taken when the union of two mutually exclusive events are considered. It can be shown that the measure of affinity defined in (4.2.1) is uniquely determined by these postulates, with the help of the following lemmas.

LEMMA 4.2.1: Let

(4.2.3) $\quad g(x,y) = \rho_2(x,1-x:y,1-y) -1, \quad x,y \in I = [0,1]$

Then,

(4.2.4) $g(x,y) = g(1-x, 1-y)$, $x, y \in I$

Proof: From postulate $C2$ for $n=3$ one gets

(4.2.5) $\rho_3(p_1, p_2, p_3 : q_1, q_2, q_3) = \rho_3(p_2, p_1, p_3 : q_2, q_1, q_3)$, $p_1+p_2+p_3 = q_1+q_2+q_3 = 1$

From $C1$ one gets

(4.2.6) $\rho_2(p_1+p_2, p_3 : q_1+q_2, q_3) + (p_1+p_2)^{\frac{1}{2}}(q_1+q_2)^{\frac{1}{2}} \rho_2(p_1/(p_1+p_2), p_2/(p_1+p_2) :$

$q_1/(q_1+q_1), q_2/(q_1+q_2)) = \rho_2(p_2+p_1, p_3 : q_2+q_1, q_3) + (p_2+p_1)^{\frac{1}{2}}(q_2+q_1)^{\frac{1}{2}}$

$\rho_2(p_2/(p_2+p_1), p_1/(p_2+p_1) : q_2/(q_2+q_1), q_1/(q_2+q_1))$

Now by cancelling the first terms on both sides of (4.2.6) the lemma is proved. Also by taking $x=y=0$ we have

(4.2.7) $g(0,0) = g(1,1)$

LEMMA 4.2.2: $g(x,y)$ satisfies the functional equation

(4.2.8) $g(x,y) + (1-x)^{\frac{1}{2}}(1-y)^{\frac{1}{2}} g(u/(1-x), v/(1-y)) = g(u,v) + (1-u)^{\frac{1}{2}}(1-v)^{\frac{1}{2}}$

$g(x/(1-u), y/(1-v))$, $x, y, u, v \in [0, 1[$, $x+u, y+v \in I$

Proof: From $C2$ for $n = 3$,

(4.2.9) $\rho_3(p_1, p_2, p_3 : q_1, q_2, q_3) = \rho_3(p_3, p_2, p_1 : q_3, q_2, q_1)$

From (4.2.9), $C1$ and lemma 4.2.1, we have

(4.2.10) $g(p_3, q_3) + (p_1+p_2)^{\frac{1}{2}}(q_1+q_2)^{\frac{1}{2}} g(p_1/(p_1+p_2), q_1/(q_1+q_2)) = g(p_1, q_1)$

$+ (p_2+p_3)^{\frac{1}{2}}(q_2+q_3)^{\frac{1}{2}} g(p_3/(p_2+p_3), q_3/(q_2+q_3))$

Since $p_1+p_2+p_3 = 1 = q_1+q_2+q_3$, (4.2.10) reduces to the following.

(4.2.11) $g(p_3, q_3) + (1-p_3)^{\frac{1}{2}}(1-q_3)^{\frac{1}{2}} g(p_1/(1-p_3), q_1/(1-q_3))$

$= g(p_1, q_1) + (1-p_1)^{\frac{1}{2}}(1-q_1)^{\frac{1}{2}} g(p_3/(1-p_1), q_3/(1-q_1))$

Now by putting $p_3=x$, $q_3=y$, $p_1=u$ and $q_1=v$, lemma 4.2.2 is proved. When p_1 and lemma 4.2.1 are used the conditions $p_1 \neq 1$, $q_1 \neq 1$, $p_3 \neq 1$, $q_3 \neq 1$ are automatically satisfied. Now it will be shown that any function satisfying the functional equation (4.2.8) is Matusita's measure of affinity. That is, this can also be taken as an alternate definition for affinity.

THEOREM 4.2.1: $\rho_n(p_1, \ldots, p_n : q_1, \ldots, q_n) = \sum_{i=1}^{n} (p_i q_i)^{\frac{1}{2}}$
is the only function ρ_n satisfying the postulates $C1, C2$ and $C3$. In other words, $C1, C2$ and $C3$ uniquely determine ρ_n as $\Sigma(p_i q_i)^{\frac{1}{2}}$.

The Concept of Affinity and Its Characterizations

Proof: This theorem will be proved by showing that $g(x,y)$ defined in (4.2.4) is of the form $x^{\frac{1}{2}}y^{\frac{1}{2}} + (1-x)^{\frac{1}{2}}(1-y)^{\frac{1}{2}} - 1$. Putting $u/(1-x) = p$, $v/(1-y) = q$, $1-x = r$, $1-y = s$ in (4.2.8) one gets

(4.2.12) $\quad g(r,s) + r^{\frac{1}{2}}s^{\frac{1}{2}}g(p,q) = g(pr,qs) + (1-pr)^{\frac{1}{2}}(1-qs)^{\frac{1}{2}}g((1-r)/(1-pr),$

$(1-s)/(1-qs))$, $r,s \in]0,1]$, $p,q \in I$, $pr \neq 1$, $qs \neq 1$

Let

(4.2.13) $\quad f(p,q,r,s) = g(r,s) + [r^{\frac{1}{2}}s^{\frac{1}{2}} + (1-r)^{\frac{1}{2}}(1-s)^{\frac{1}{2}}]g(p,q)$, $p,q,r,s \in]0,1[$

We will show that $f(p,q,r,s)$ is symmetric in pairs (p,r) and (q,s). Now by using (4.2.12) successively we get

(4.2.14) $\quad f(p,q,r,s) = g(pr,qs) + (1-pr)^{\frac{1}{2}}(1-qs)^{\frac{1}{2}}g((1-r)/(1-pr),(1-s)/(1-qs))$

$\qquad + (1-r)^{\frac{1}{2}}(1-s)^{\frac{1}{2}}g(p,q)$

(4.2.15) $\qquad = g(pr,qs) + (1-pr)^{\frac{1}{2}}(1-qs)^{\frac{1}{2}}\{g((1-r)/(1-pr),(1-s)/(1-qs))$

$\qquad + [(1-r)/(1-pr)]^{\frac{1}{2}}[(1-s)/(1-qs)]^{\frac{1}{2}}g(p,q)\}$

(4.2.16) $\qquad = g(pr,qs) + (1-pr)^{\frac{1}{2}}(1-qs)^{\frac{1}{2}}\{g(p(1-r)/(1-pr),q(1-s)/(1-qs))$

$\qquad + [(1-p)/(1-pr)]^{\frac{1}{2}}[(1-q)/(1-qs)]^{\frac{1}{2}}g(r,s)\}$

Now by using the result $g(r,s) = g(1-r,1-s)$ we have

(4.2.17) $\quad f(p,q,r,s) = g(pr,qs) + (1-pr)^{\frac{1}{2}}(1-qs)^{\frac{1}{2}}\{g((1-p)/(1-pr),(1-q)/(1-qs))$

$\qquad + [(1-p)/(1-pr)]^{\frac{1}{2}}[(1-q)/(1-qs)]^{\frac{1}{2}}g(r,s)\}$

Now comparing (4.2.15) and (4.2.17) we see that $f(p,q,r,s)$ is symmetric in (p,r) and (q,s). Therefore, from (4.2.12) and (4.2.13),

(4.2.18) $\quad f(p,q,r,s) = g(p,q) + [p^{\frac{1}{2}}q^{\frac{1}{2}} + (1-p)^{\frac{1}{2}}(1-q)^{\frac{1}{2}}]g(r,s)$

$\qquad = g(r,s) + [r^{\frac{1}{2}}s^{\frac{1}{2}} + (1-r)^{\frac{1}{2}}(1-s)^{\frac{1}{2}}]g(p,q)$

That is,

(4.2.19) $\quad g(r,s) = [r^{\frac{1}{2}}s^{\frac{1}{2}} + (1-r)^{\frac{1}{2}}(1-s)^{\frac{1}{2}} -1]g(p,q)/[p^{\frac{1}{2}}q^{\frac{1}{2}} + (1-p)^{\frac{1}{2}}(1-q)^{\frac{1}{2}} -1]$

Now p and q are at our choice subject to the condition that $p,q \in]0,1[$. By using the condition given by postulate $C3$ the second factor is cancelled and (4.2.19) yields

(4.2.20) $\quad g(r,s) = r^{\frac{1}{2}}s^{\frac{1}{2}} + (1-r)^{\frac{1}{2}}(1-s)^{\frac{1}{2}} - 1$, $r,s \in]0,1[$

That is,

(4.2.21) $\quad \rho_2(r,s) = r^{\frac{1}{2}}s^{\frac{1}{2}} + (1-r)^{\frac{1}{2}}(1-s)^{\frac{1}{2}}$

It may be noticed that $\rho_2(r,s)=1$ when $r=s$. This agrees with the convention that the measure of affinity is maximum when the vectors

$(p_1,\ldots,p_n)$ and $(q_1,\ldots,q_n)$ coincide. Now we will extend (4.2.19) to the closed interval $[0,1]$. To this end we have to show that (4.2.19) holds for $g(0,y)$ and $g(x,0)$. Since $g(x,y)=g(1-x,1-y)$ the other points follow automatically. By putting $p=q=1$ in (4.2.12) we get

(4.2.22) $[r^{\frac{1}{2}}s^{\frac{1}{2}} - (1-r)^{\frac{1}{2}}(1-s)^{\frac{1}{2}}]g(1,1) = 0$, for all $r,s \in]0,1[$

That is,

(4.2.23) $g(1,1) = 0 = g(0,0)$

By putting $p=0$ and $s=1$ in (4.2.12) we get

(4.2.24) $(1-r)^{\frac{1}{2}}g(0,q) = g(r,1) - (1-q)^{\frac{1}{2}}g(1-r,0) = [1-(1-q)^{\frac{1}{2}}]g(r,1)$,

$r \in]0,1]$, $q \in [0,1[$.

By putting $p=0$ and $s=\frac{1}{2}$ in (4.2.12) we get

(4.2.25) $g(r,\frac{1}{2}) + (r/2)^{\frac{1}{2}}g(0,q) = g(0,q/2) + (1-q/2)^{\frac{1}{2}}g(1-r, 1/(2-q))$

for $r \in]0,1]$, $q \in I$. Now by putting $r=\frac{1}{2}$ and using the value of $g(x,y)$ given in (4.2.20) for points inside $]0,1[$ we get

(4.2.26) $(1/2)g(0,q) = g(0,q/2) + (1-q/2)^{\frac{1}{2}}[1/(4-2q)^{\frac{1}{2}} + (1-q)^{\frac{1}{2}}/(4-2q)^{\frac{1}{2}} -1]$

for $q \in [0,1[$. Now substituting in (4.2.26) the values of $g(0,q)$ and $g(0,q/2)$ from (4.2.24) to (4.2.26) we get

(4.2.27) $g(r,1)[\{1-(1-q)^{\frac{1}{2}}\}/\{2(1-r^{\frac{1}{2}})\} - \{1-(1-q/2)^{\frac{1}{2}}\}/(1-r^{\frac{1}{2}})]$

$= (1-q/2)^{\frac{1}{2}}\{1/(4-2q)^{\frac{1}{2}} + (1-q)^{\frac{1}{2}}/(4-2q)^{\frac{1}{2}}\}$

That is,

(4.2.28) $g(r,1) = r^{\frac{1}{2}} - 1$, $r \in]0,1]$

and

(4.2.29) $g(0,q) = (1-q)^{\frac{1}{2}} - 1$, $q \in [0,1[$

Now we have $g(0,y)$ for $y \in [0,1[$ and $g(x,1)$ for $x \in]0,1]$. We already have $g(x,y)=g(1-x,1-y)$. So we need only $g(0,1)$ more. Now putting $q=1$ and $r=\frac{1}{2}$ in (4.2.25) and using (4.2.28) and (4.2.29) one gets

(4.2.30) $g(0,1) = -1$

This completes the proof that

(4.2.31) $g(r,s) = r^{\frac{1}{2}}s^{\frac{1}{2}} + (1-r)^{\frac{1}{2}}(1-s)^{\frac{1}{2}} - 1$, $r,s \in I$

Now by using the recurrence relation in postulate $C1$ successively we have

(4.2.32) $\rho_n(p_1,\ldots,p_n;q_1,\ldots,q_n) - 1 = \rho_{n-1}(p_1+p_2,p_3,\ldots,p_n;q_1+q_2,q_3,\ldots,q_n)$

$- 1 + (p_1+p_2)^{\frac{1}{2}}(q_1+q_2)^{\frac{1}{2}}\{\rho_2(p_1/(p_1+p_2),p_2/(p_1+p_2);q_1/(q_1+q_2),$

The Concept of Affinity and Its Characterizations

$$(4.2.33) \quad = \Sigma_{i=2}^{n} r_i^{\frac{1}{2}} s_i^{\frac{1}{2}} \{\rho_2(p_i/r_i, 1-p_i/r_i: q_i/s_i, 1-q_i/s_i) - 1\}$$

where $r_i = p_1 + \ldots + p_i$, $s_i = q_1 + \ldots + q_i$;

$$(4.2.34) \quad = \Sigma_{i=2}^{n} r_i^{\frac{1}{2}} s_i^{\frac{1}{2}} \{(p_i q_i / r_i s_i)^{\frac{1}{2}} + (1-p_i/r_i)^{\frac{1}{2}}(1-q_i/s_i)^{\frac{1}{2}} - 1\}$$

$$(4.2.35) \quad = \Sigma_{i=2}^{n} p_i^{\frac{1}{2}} q_i^{\frac{1}{2}} + \Sigma_{i=2}^{n} r_{i-1}^{\frac{1}{2}} s_{i-1}^{\frac{1}{2}} - \Sigma_{i=2}^{n} r_i^{\frac{1}{2}} s_i^{\frac{1}{2}}$$

$$(4.2.36) \quad = \Sigma_{i=2}^{n} p_i^{\frac{1}{2}} q_i^{\frac{1}{2}} + r_1^{\frac{1}{2}} s_1^{\frac{1}{2}} - r_n^{\frac{1}{2}} s_n^{\frac{1}{2}}$$

$$(4.2.37) \quad = \Sigma_{i=1}^{n} p_i^{\frac{1}{2}} q_i^{\frac{1}{2}} - 1$$

That is,

$$(4.2.38) \quad \rho_n = \Sigma_{i=1}^{n} (p_i q_i)^{\frac{1}{2}}$$

This completes the proof.

Now a characterization theorem will be given based on a maximization principle. For convenience we will characterize Matusita's measure of squared distance between the distributions which is defined as

$$(4.2.39) \quad d_n = \Sigma_{i=1}^{n} (p_i^{\frac{1}{2}} - q_i^{\frac{1}{2}})^2 = 2(1 - \Sigma p_i^{\frac{1}{2}} q_i^{\frac{1}{2}})$$

This characterization of d_n will also lead to a characterization of affinity. Let $K_n(p_1,\ldots,p_n; q_1,\ldots,q_n)$, $p_i, q_i > 0$, $i = 1, 2, \ldots, n$, $\Sigma p_i = 1 = \Sigma q_i$, $n \geq 3$ be any function of p_i's and q_i's satisfying the following postulates.

D1. *Structure*: $K_n = \Sigma_{i=1}^{n} p_i^{\frac{1}{2}} (f(p_i) - f(q_i))$ for some function $f(x)$;
D2. *Non-negativity*: $K_n \geq 0$, for all $n \geq 3$;
D3. *Normalization*: $K_2(1/4, 3/4: 3/4, 1/4) = 4 \sin^2(\pi/12)$.

Here the postulates D1 and D2 give some sort of a maximization principle. That is, $\Sigma p_i^{\frac{1}{2}} f(q_i)$ has the maximum value when $q_i = p_i$ for all i.

THEOREM 4.2.2: K_n satisfying the postulates D1, D2 and D3 is uniquely determined as

$$(4.2.40) \quad K_n = d_n = 2(1 - \Sigma_{i=1}^{n} p_i^{\frac{1}{2}} q_i^{\frac{1}{2}})$$

The proof is omitted because a general case is already discussed in chapter 2, section 2.1.

A measure of affinity among r discrete distributions is defined as

$$(4.2.41) \quad A = \Sigma_{i=1}^{n} (p_{1i} \cdots p_{ri})^{1/r}$$

where $(p_{j1}, p_{j2}, \ldots, p_{jn})$, $j = 1, 2, \ldots, r$ denote the r discrete distributions and a continuous analogue can also be defined in a similar fashion. For a characterization of this measure see Kaufman and Mathai (1973).

4.3 A MEASURE OF DISCREPANCY

In statistical tests of goodness-of-fit a statistic usually used is Pearson's χ^2 statistic which is defined as follows.

(4.3.1) $\quad \chi^2 = \sum_{i=1}^{k} \{(n_i - np_i)^2 / np_i\}$

where n_i and np_i are the observed and expected frequencies corresponding to the i-th group respectively and where $n = \Sigma n_i$ and p_i is the probability of getting an observation in the i-th group. χ^2 can be simplified to the form,

(4.3.2) $\quad \chi^2 = n\{\sum_{i=1}^{k}(q_i^2/p_i) - 1\}, \quad q_i = n_i/n$

(a) DEFINITION: The quantity $\Sigma q_i^2 / p_i - 1$ can be considered to be a measure of discrepancy between the two discrete populations $(p_1, \ldots, p_k)$ and $(q_1, \ldots, q_k)$ because this quantity which is quite useful in tests of statistical hypotheses can be considered to be a measure of separation of the two distributions even though $\Sigma q_i^2 / p_i - 1$ is not a distance measure.

(b) CHARACTERIZATIONS: Two different characterizations will be considered here. One is based on a recursivity property and the other is based on the assumption of a particular structure for the measure of discrepancy. Let $D_k(q_1, \ldots, q_k : p_1, \ldots, p_k)$ for $p_i, q_i \geq 0$, $i = 1, 2, \ldots, k$, $\Sigma p_i = 1 = \Sigma q_i$ satisfy the following postulates.

E1. $D_k(q_1, \ldots, q_k : p_1, \ldots, p_k) = D_{k-1}(q_1 + q_2, q_3, \ldots, q_k : p_1 + p_2, p_3, \ldots, p_k)$
$\quad + (q_1 + q_2)^2 (p_1 + p_2)^{-1} D_2(q_1/(q_1+q_2), q_2/(q_1+q_2) : p_1/(p_1+p_2), p_2/(p_1+p_2))$
$\quad$ for $p_1 + p_2$, $q_1 + q_2 > 0$ and for all $n \geq 3$;

E2. $D_3(q_1, q_2, q_3 : p_1, p_2, p_3)$ is symmetric in pairs $\{p_i, q_i\}$, $i = 1, 2, 3$;

E3. $D_2(1/2, 1/2 : 1/4, 3/4) = 1/3$.

THEOREM 4.3.1: The postulates E1, E2 and E3 uniquely determine $D_k(q_1, \ldots, q_k : p_1, \ldots, p_k)$ for $p_i, q_i \geq 0$, $i = 1, 2, \ldots, k$, $\Sigma p_i = \Sigma q_i = 1$ as $\Sigma q_i^2 / p_i - 1$.

Since the proof is similar to the proof of theorem 4.2.1 it is omitted.

Another interesting characterization can be given by assuming a particular structure. Let $D_k(q_1, \ldots, q_k : p_1, \ldots, p_k)$ be a function of

A Measure of Discrepancy

$q_1,\ldots,q_k$ and $p_1,\ldots,p_k$ satisfying the following postulates.

F1. $D_k(q_1,\ldots,q_k;p_1,\ldots,p_k)$ is of the form $\Sigma q_i f(q_i)/f(p_i) - 1$ for $p_i, q_i > 0$, $i = 1,2,\ldots,k$, $\Sigma p_i = 1 = \Sigma q_i$ for fixed $k \geq 3$ where $f(p) > 0$ for all $p \in\ ]0,1[\ $;

F2. $D_k(q_1,\ldots,q_k;p_1,\ldots,p_k) \geq 0$;

F3. $D_k(1/8, 3/8, 1/2(k-2),\ldots,1/2(k-2); 3/8, 1/8, 1/2(k-2),\ldots,1/2(k-2)) = 2/3$.

Then the following result can be shown which is stated here without proof. The proof is similar to the proof of theorem 4.2.2.

LEMMA 4.3.1: The function D_k satisfying the postulates F1 to F3 has exactly two forms given by either $\Sigma(q_i^2/p_i) - 1$ or $\Sigma(p_i^2/q_i) - 1$.

Now a characterization theorem can be stated in the following form.

THEOREM 4.3.2: The function D_k satisfying the postulates F1, F2, F3, and the postulate,

F4. For fixed $(q_1,\ldots,q_k)$ if $p_i \to 0$ for any i then $D_k \to \infty$, uniquely determine D_k as $\Sigma(q_i^2/p_i) - 1$ for $p_i, q_i > 0$, $j = 1,2,\ldots,k$, $\Sigma p_i = 1 = \Sigma q_i$ and fixed $k \geq 3$.

4.4 THE CONCEPT OF DISPERSION

A measure of dispersion or scatter in a statistical population is an important and useful concept. It is pointed out in Mathai (1967) that a variety of statistical problems of estimation and tests of hypotheses can be considered to be the study of properly defined measures of dispersion. Here we will consider only one type of dispersion, namely, the scatter of a stochastic variable from a point of reference. Other types of scatter are total range of values the stochastic variables assume with non-zero probabilities and the scatter among individual values assumed by a discrete stochastic variable.

(a) DEFINITION: If a statistical population is designated by a univariate stochastic variable X' and if $X = X' - m$ where m is a fixed point of reference, a measure of dispersion in X' from m, can be defined by the following metric D satisfying the following axioms.

G1. $D(X) \geq 0$, $D(X) = 0 \iff X = 0$ almost surely;

G2. $D(aX) = |a|\ D(X)$, where a is a scalar quantity;

G3. $D(X + Y) \leq D(X) + D(Y)$ where $Y = Y' - m$ and Y' is another stochastic variable;

G4. $D(X) = 1$ if $|X| = 1$ almost surely.

A $D(X)$ satisfying G1 to G4 will be called a measure of dispersion in X. Some examples are the following where E denotes the operator *mathematical expectation*.

(4.4.1) $\quad D_r = \{E|X|^r\}^{1/r}$ for fixed $r \geqslant 1$.

(4.4.2) $\quad D_{r,k} = \{\Sigma_{i=1}^{k} c_i D_i^r\}^{1/r}$, for fixed r and k, $r \geqslant 1$, $c_i \geqslant 0$, $\Sigma c_i = 1$, $k = 1, 2, \ldots$

When X is a discrete stochastic variable taking the values $x_1 = x_1' - m, \ldots, x_n = x_n' - m$ with non-zero probabilities $p_1, \ldots, p_n$, $\Sigma p_i = 1$, then the following quantities will also satisfy axioms $G1$ to $G4$.

(4.4.3) $\quad d_r = \{\Sigma_{i=1}^{n} p_i |x_i|^r\}^{1/r}$, for fixed $r \geqslant 1$, $p_i > 0$, $\Sigma p_i = 1$

(4.4.4) $\quad d_{r,k} = \{\Sigma_{i=1}^{k} b_i d_i^r\}^{1/r}$, for fixed r and k, $r \geqslant 1$, $b_i \geqslant 0$, $\Sigma b_i = 1$, $k = 1, 2, \ldots$

(4.4.5) $\quad d = \max_{i} |x_i|$

When $p_1 = \ldots = p_n = 1/n$ and $m = \bar{x} = (x_1 + \ldots + x_n)/n$ then d_r in (4.4.3) for $r = 1, 2$ defines the usual measures of mean absolute deviation and standard deviation in a given data respectively. D_r^r of (4.4.1) are the various absolute moments which are usually used in statistical literature. Postulates $G1$ to $G4$ are also intuitively justifiable because they are some of the properties of measures of scatter, such as standard deviation, which are used in statistical literature.

(b) *JOINT, MULTIVARIATE AND GENERALIZED DISPERSIONS:* If (X', Y') is a pair of stochastic variables having a joint distribution and if (m_1, m_2) is a point of reference then the joint dispersion $C(X, Y)$ in $(X, Y) = (X' - m_1, Y' - m_2)$ can be defined by the following axioms.

H1. $C(X, Y) = C(Y, X)$;

H2. $C(aX, Y) = a\, C(X, Y)$ where a is a scalar quantity;

H3. $C(X+Z, Y) = C(X, Y) + C(Z, Y)$ where $Z = Z' - m_3$ and Z' is another stochastic variable having a joint distribution with (X, Y);

H4. $C(X, X) = \{D(X)\}^2$ where $D(X)$ is a univariate measure of dispersion defined by the postulates $G1$ to $G4$.

An example of a measure of joint dispersion is the concept of covariance in statistical literature which is defined in section 4.1. Axioms $H1$ to $H4$ are also desirable properties of a measure of joint dispersion from a statistical point of view because these are satisfied by covariance which is already used in statistical literature as a measure of joint dispersion. Now consider a set of k stochastic variables having a joint distribution. Let C_{ij} denote a measure of joint dispersion in the i-th and j-th variates then the matrix $C =$

The Concept of Dispersion

(C_{ij}) can be taken as a multivariate measure of joint dispersion in the k variates under consideration. A measure of generalized dispersion is a scalar quantity which is given as a multivariate analogue of a univariate measure of dispersion. Hence a generalized dispersion can be defined as any *norm* of the matrix C satisfying the usual axioms for a norm, namely,

I1. $||C|| > 0$, $||C|| = 0 \iff C = \phi$ where ϕ is a null matrix;
I2. $||aC|| = |a| \, ||C||$ where a is a scalar quantity;
I3. $||C+D|| \leq ||C|| + ||D||$ where D is another matrix defined in a similar fashion;
I4. $||CD|| \leq ||C|| \, ||D||$.

Thus when C is the usual variance-covariance matrix then any norm of C can be called a generalized variance. Some examples are the following.

(4.4.6) $\quad V_1 = \max_i \Sigma_j |C_{ij}|$.

(4.4.7) $\quad V_2 = $ largest eigenvalue of C.

(4.4.8) $\quad V_3 = k \max_{ij} |C_{ij}|$.

(4.4.9) $\quad V_4 = (\Sigma |C_{ij}|^2)^{\frac{1}{2}}$.

(c) *DISTANCE BETWEEN TWO POPULATIONS:* A concept of distance between two statistical populations is quite useful in problems of classification, discrimination, tests of hypotheses and several other problems of statistical decision making. If two statistical populations are designated by the stochastic variables X and Y and if $X=Y$ almost surely then the distance between X and Y may be taken as zero. Hence $e = X-Y$ may be taken as an error in taking Y for X and any measure of dispersion in e, namely $D(e)$ satisfying the axioms $G1$ to $G4$, can be defined as a measure of distance between X and Y. If X and Y are in different units of measurements then instead of $X-Y$ one may take $X/D(X) - Y/D(Y)$ as e where $D(X)$ and $D(Y)$ are some fixed measures of dispersion in X and Y.

Consider two multinomial populations given by the probability distributions $(p_1', \ldots, p_k')$ and $(q_1', \ldots, q_k')$ with $p_i', q_i' > 0$ for all i, $\Sigma p_i' = 1 = \Sigma q_i'$. If possible let us represent them as points A and B on a hypersurface $\Sigma |x_i|^s = 1$ for some fixed $s \geq 1$. In this case the two points are $(p_1, \ldots, p_k)$ and $(q_1, \ldots, q_k)$ where $p_i^s = p_i'$, $q_i^s = q_i'$ for all i. In this representation if the two points concide then the distance between them may be taken as zero. Hence a measure of dispersion $D(A-B)$ satisfying axioms $G1$ to $G4$ may be taken as a measure of distance between the two multinomial distributions. Some examples are the following:

(4.4.10) $\quad P_1 = \{\Sigma | p_i'^{1/s} - q_i'^{1/s} |^r\}^{1/r}$ for fixed $\quad r \geq 1, s \geq 1$

(4.4.11) $\quad P_2 = \max_i |p_i'^{1/s} - q_i'^{1/s}|$ for fixed $\quad s \geq 1$

The continuous analogue of P_1 is,

(4.4.12) $\quad P_3 = \{\int |[p(x)]^{1/s} - [q(x)]^{1/s}|^r \, dx\}^{1/r}$

When $r=s=1$, (4.4.12) is often called *Kolmogorov's variational distance*. In (4.4.10) if $r=s=m$ then P_1^m is the measure called *Jeffreys' Invariant* I_m discussed in Jeffreys (1961). If $r=s=2$ then P_1 is Matusita's measure of distance discussed in section 4.2. $Cos^{-1}(1-I_2/2)$ is Bhattacharyya's measure of distance (Adke(1958)) where I_m is Jeffreys' Invariant. Thus a class of measures of distances between two multinomial populations, which are used in statistical literature, are available from P_1. For more properties of some of these measures see the problems at the end of this chapter.

(d) THE PRINCIPLE OF MINIMUM DISPERSION: Let $x=(x_1,\ldots,x_n)$ be an observation vector and let $y(\theta) = (y_1(\theta),\ldots,y_n(\theta))$ be a model set up for x. In other words let $y(\theta)$ be a hypothetical vector designed for x by a statistical hypothesis, where θ is defined in some parameter space. A good $y(\theta)$ for x is the one which will minimize a measure of dispersion $D(x-y(\theta))$ with respect to θ in some sense. So a *best* estimator for θ is that θ, if it exists, which minimizes a $D(x-y(\theta))$. This principle may be called the principle of minimum dispersion. It can be shown that a good number of statistical techniques of estimation are minimizations of different measures of dispersion. For example, consider the following measures of dispersion.

(4.4.13) $\quad m_r = \{(1/n)\Sigma |x_i - y_i(\theta)|^r\}^{1/r}$

(4.4.14) $\quad m = \max_i |x_i - y_i(\theta)|$

Minimization of m_r for $r=2$ leads to the principle of least squares and the minimization of m leads to a mimimax principle of estimation. If a statistician makes a decision in a situation designated by θ the loss or gain (may be economic) in making this decision t about θ is in some sense a measure of deviation from the correct decision and it may be denoted by $t-\theta$. A measure of dispersion $D(t-\theta)$ may be called his risk in taking the decision t. An example of $D(t-\theta)$ is $\{E|t-\theta|^r\}^{1/r}$ for fixed $r \geq 1$. Minimization of this risk leads to

The Concept of Dispersion

minimum risk estimates. Also minimum value of a measure of dispersion can be taken as a test statistic for the purpose of testing statistical hypotheses. For example see problems at the end of this chapter.

4.5 OTHER MEASURES

There are several other measures which are fundamental in probability and statistics. Among measures of association among stochastic variables, the important measures, which are frequently used in statistical analysis, are linear, multiple, partial and serial correlation coefficients, inter-class and intra-class correlation coefficients, correlation ratio, square contingency and Pearson's coefficient of contingency as measures of association in contingency tables, and so on. As measures of skewness in distributions, $E(X-EX)^{2r+1}$, $r = 0,1,...$ are usually used. Kurtosis of distributions is often measured by $\mu_4/\mu_2^2 - 3$ where μ_4 and μ_2 are the 4-th and 2-nd central moments respectively. Apart from the measures of dispersion discussed in section 4.4 the extend of variation in a given data or in a distribution is often measured by range, inter-quartile range, median and percentile points. Dispersion among the observations in a given sample is often measured by Gini's mean difference, namely, $\Sigma_{ji}|x_i - x_j|/\{n(n-1)\}$. Gini also defines a concentration ratio as

$$G^* = (1/2)\Sigma_{i=1}^n \Sigma_{j=1}^n |x_i y_j - x_j y_i| = (1/2)\Sigma_{i=1}^n \Sigma_{j=1}^n x_i x_j |y_i/x_i - y_j/x_j|$$

Gini has defined an index of diversity by the expression $1 - \Sigma p_i^2$ where $(p_1,...,p_n)$ is a discrete distribution.

Another measure which is frequently used in statistical inference is Fisher's measure of information about a parameter θ available from a sample, which is defined as $E[\frac{\partial \log L}{\partial \theta}]^2$ where L is the likelihood function and E stands for 'mathematical expectation'.

Another measure is the amount of information contained in one random variable about another random variable which is defined as

$$J(\xi,\eta) = \Sigma_{i=1}^s \Sigma_{j=1}^t r_{ij} \log(r_{ij}/p_i q_j)$$

where ξ and η are random variables taking the values $x_1,...,x_s$ and $y_1,...,y_t$ with probabilities $p_1,...,p_s$ and $q_1,...,q_t$ respectively and (x_i,y_j) with joint probability r_{ij} for $i = 1,...,s$, $j=1,...,t$. For a general class of coefficient of divergence of one distribution from another see Ali and Silvey (1966). These and other such measures are available from standard books on statistics and probability.

4.6 APPLICATIONS

It is pointed out in section *4.4(c)* that in statistical estimation problems the different methods of estimation used in the literature can be looked upon as the minimizations of one or the other measure of dispersion. Since the concept of affinity may be interpreted as the angular distance between two distributions this concept measures the closeness of the distributions in some sense. For the applications of affinity in point and interval estimation problems see Matusita (1954,1955,1961). It is shown in George and Mathai (1974) that this concept of affinity is quite useful in classifying populations according to some socio-anthropological characteristics. Some applications of Gini's measures to problems in Ecology may be found in Bhargava and Uppuluri (1971). Some uses and interpretations of distance functions in statistics may be seen from Rao (1949,1954).

4.7 OPEN PROBLEMS

It is desirable to write down the various sets of postulates which will characterize the concepts of covariance, variance, distance, dispersion etc. Characterizations of the various types of correlations are open problems. In this chapter we have given the various characterization theorems when the populations are discrete but their continuous analogues are still open problems. A proper physical interpretation of affinity among a number of distributions is yet to be found. Analogous to Pearson's measure of discrepancy one may investigate measures and their characterizations associated with multiway classifications. Only very little has been done so far in the axiomatic characterizations of the various measures of dispersion. This will lead to axiomatic foundations for a number of theories of statistical estimation procedures.

Problems *4.1*, *4.2*, *4.5*, *4.6*, *4.9*, *4.10*, *4.11* and *4.12* listed in the exercises are taken from Mathai (1967), Mathai and Rathie (1971,1971a) and Kaufman, Mathai and Rathie(1971).

EXERCISES

4.1 Show that the covariance defined in *(4.1.1)* can be characterized by using the following set of five postulates when $n=2$.
(1) $f(x_1+z_1, x_2+z_2 : y_1, y_2) = f(x_1, x_2 : y_1, y_2) + f(z_1, z_2 : y_1, y_2)$ for $x_1, x_2, y_1, y_2, z_1, z_2 \in (-\infty, \infty)$;

Exercises

(2) $f(x_1,x_2:y_1,y_2)$ is a continuous function of its variables;
(3) $f(x_1,x_2:y_1,y_2) = f(y_1,y_2:x_1,x_2)$;
(4) $f(x,x:y_1,y_2) = 0$;
(5) $f(1,-1:1,-1) = 1$.

4.2 Show that the covariance between two stochastic variables defined in (4.1.2) with $EX=EY=0$ can be characterized by the following postulates once it is assumed that $\int_{-\infty}^{\infty}\int_{-\infty}^{\infty} g(x,y)dF(x,y) = cov(X,Y)$ where $F(x,y)$ is the joint distribution function and $g(x,y)$ satisfies the postulates,

(1) $g(x+z,y) = g(x,y) + g(z,y)$;
(2) $g(x,y) = g(y,x)$;
(3) $g(x,y)$ is continuous;
(4) $g(1,1) = 1$.

4.3 Show that the variance in (4.1.12) is characterized by the postulate G1 and the postulate $V(ax_1,\ldots,ax_n) = a^2 V(x_1,\ldots,x_n)$.

4.4 Show that the linear correlation coefficient $r(x_1,\ldots,x_n:y_1,\ldots,y_n) = \Sigma(x_i-\bar{x})(y_i-\bar{y})/\{\Sigma(x_i-\bar{x})^2 \Sigma(y_i-\bar{y})^2\}^{1/2}$ can be characterized by using theorems 4.1.1 and 4.1.2 and by assuming a structure $r(x_1,\ldots,x_n:y_1,\ldots,y_n) = H(x_1,\ldots,x_n:y_1,\ldots,y_n)/\{G(x_1,\ldots,x_n)G(y_1,\ldots,y_n)\}$.

4.5 Give characterization theorems for Bhattacharyya's measure of distance $\Phi = \cos^{-1}(\Sigma p_i^{\frac{1}{2}} q_i^{\frac{1}{2}})$ and Matusita's measure of distance $\Sigma(p_i^{\frac{1}{2}} - q_i^{\frac{1}{2}})^2$ by using theorem 4.2.1.

4.6 Let $M_{n,r,s} = \Sigma_{j=0}^{2r} \binom{2r}{j}(-1)^j \Phi_n(p_1,\ldots,p_n:q_1,\ldots,q_n)$ for fixed $r,s \geq 1$, $s \neq 2r$, $p_i, q_i \geq 0$, $i=1,\ldots,n$; $\Sigma p_i = 1 = \Sigma q_i$ and let $M_{n,r,s} = 2 + \Sigma_{j=1}^{2r-1} \binom{2r}{j}(-1)^j \Phi_n(p_1,\ldots,p_n:q_1,\ldots,q_n)$ when $s = 2r$. Let $\Phi_n(p_1,\ldots,p_n:q_1,\ldots,q_n)$ satisfy the following postulates.

(1) $\Phi_n(p_1,\ldots,p_n:q_1,\ldots,q_n) = \Phi_{n-1}(p_1+p_2,p_3,\ldots,p_n:q_1+q_2,q_3,\ldots,q_n)$
$+ (p_1+p_2)^{j/s}(q_1+q_2)^{(2r-j)/s}\{\Phi_2(p_1/(p_1+p_2),p_2/(p_1+p_2):q_1/(q_1+q_2),q_2/(q_1+q_2)) - 1\}$ for all $n \geq 3$, p_1+p_2, $q_1+q_2 > 0$, for fixed $s \geq 1$, $s \neq 2r$ or $s = 2r$ and $j \neq 0$, $j \neq 2r$;

(2) Φ_3 is symmetric in $\{p_i,q_i\}$, $i=1,2,3$;

(3) $\Phi_2(1/4,3/4:3/4,1/4) = (3^{j/s} + 3^{(2r-j)/s})2^{-4r/s} - 1$.

Then show that $M_{n,r,s}$ is uniquely determined as the generalized measure of distance,

$$M_{n,r,s} = \Sigma_{i=1}^{n}(p_i^{1/s} - q_i^{1/s})^{2r}, \quad s \geq 1, \, r \geq 1,$$

between the two populations $(p_1,\ldots,p_n)$ and $(q_1,\ldots,q_n)$.

4.7 Give a characterization to the measure $\sum_{i=1}^{n}(p_{1i}^{a_1} p_{2i}^{a_2} \ldots p_{ri}^{a_r})$ where p_{ji}, $j=1,2,\ldots,r$ and $i=1,2,\ldots,n$ denote r discrete distributions and $a_j > 0$ for $j=1,2,\ldots,r$, by using postulates similar to the ones assumed in theorem 4.2.1.

4.8 Prove theorems 4.3.1, 4.3.2 and lemma 4.3.1.

4.9 Let $L(x,\theta_1,\ldots,\theta_k)$ be a likelihood function with $(\theta_1,\ldots,\theta_k)$ ε Ω (parameter space). Consider a neighbouring point $(\theta_1+h_1,\ldots,\theta_k+h_k)$ in Ω. Let $e = L(x,\theta_1+h_1,\ldots,\theta_k+h_k) - L(x,\theta_1,\ldots,\theta_k) \simeq \Sigma h_i(\frac{\partial L}{\partial \theta_i})$. Show that the principle of maximum likelihood estimation is equivalent to a principle of minimum dispersion. That is, show that a maximum likelihood estimate also minimizes a measure of dispersion in e as defined in section 4.4.

4.10 Let $S_n(x) = \begin{cases} 0, & x < x_{(1)} \\ r/n, & x_{(r)} \leq x < x_{(r+1)} \\ 1, & x_{(n)} \leq x \end{cases}$

and let $F(x,\theta_0)$ be the population cumulative distribution function, where $x_{(1)} \leq x_{(2)} \leq \ldots \leq x_{(n)}$ are sample observations ordered according to their magnitudes. Let $e = S_n(x) - F(x,\theta_0)$. Show that W^2 and Kolmogorov goodness-of-fit statistics are in fact based on dispersion measures in e.

4.11 Consider the same e as in problem 4.10 and consider the relative deviation e/L. (a) Show that the quadratic form $Q = \Sigma_{ij} h_i h_j E(\frac{1}{L}\frac{\partial L}{\partial \theta_i})(\frac{1}{L}\frac{\partial L}{\partial \theta_j})$ is a measure of dispersion in e/L. Note: The coefficient of $h_i h_j$ in Q is Fisher's Information measure;

(b) Show that when two multivariate populations differ only in their means, Mahalanobis' measure of generalized distance between two populations is obtained by integrating $Q^{1/2}$ along a geodesic in the Riemannian space of k dimensions.

CHAPTER 5

SOME OTHER MEASURES AND INEQUALITIES

5.0 INTRODUCTION

This chapter deals with the measures, analogous to the various measures of entropies, directed divergences and inaccuracies, which are defined for sequences of non-negative numbers with the sum being equal to or less than unity. In addition, some measures useful in statistical pattern recognition and inequalities are also discussed.

The various characterization theroems are given in the papers of Aczél (1964), Aczél and Daróczy (1963), Aczél and Nath (1972), Arlotti (1970-71), Daróczy (1963), Kaufman and Rathie (1970), Nath (1968), Rathie (1970,1970a,1971b,1971c,1973,1974), Rényi (1961) and Vajda (1968). Some of these theorems are given as exercises at the end of this chapter.

5.1 MEASURES NOT INVOLVING PROBABILITIES

In this section we will define measures analogous to entropies, directed divergences and inaccuracies and their α-forms. Rényi(1961) considers the sequences of non-negative numbers $p_1,\ldots,p_n$ and $q_1, \ldots,q_n$ satisfying the conditions $0 < \Sigma_{i=1}^{n} p_i \leq 1$, $0 < \Sigma_{i=1}^{n} q_i \leq 1$, and $0 < \Sigma_{i=1}^{n} p_i + \Sigma_{i=1}^{n} q_i \leq 1$ and gives a number of characterization theorems for several functions of p's and q's which are analogous to the measures of entropy or directed divergence in discrete probability distributions. As can be expected, when the conditions on p's are relaxed so as to have $\Sigma_{i=1}^{n} p_i$ to assume any value in $]0,1]$ the problem of characterization of the function $-\Sigma_{i=1}^{n} p_i \log p_i / \Sigma_{i=1}^{n} p_i$ becomes simpler. $[-\Sigma_{i=1}^{n} p_i \log p_i]$ is Shannon's entropy when the p's form a discrete probability distribution, nemely, $p_i \geq 0$, $i=1,\ldots,n$, $\Sigma_{i=1}^{n} p_i = 1$. Rényi (1961) calls the system $(p_1,\ldots,p_n)$, $p_i \geq 0$, $i=1,\ldots,n$, $0 < \Sigma_{i=1}^{n} p_i \leq 1$ as an *incomplete distribution* and the system $(p_1,\ldots,p_n,q_1,\ldots,q_n)$, $p_i \geq 0$, $q_i \geq 0$, $0 < \Sigma_{i=1}^{n} p_i + \Sigma_{i=1}^{n} q_i \leq 1$ as the *union of two incomplete distributions*. The various results obtained in Rényi(1961) can be considered to be results in number theory and they do not seem to have much relevance to probability theory. Corresponding to an 'incomplete distribution' Rényi(1961) also talks about an *incomplete random variable* and interprets it as a quantity describing the result of an experiment depending on chance which is not always observable. If a particular event A is not observable and thus a sample space has become 'incomplete' with the sum of probabilities of observable events as $\sum_{i=1}^{n} p_i$ with $0 < \sum_{i=1}^{n} p_i \leq 1$ then evidently the probability of A, denoted by p', is $p' = 1 - \sum_{i=1}^{n} p_i$. Thus $p' + \sum_{i=1}^{n} p_i = 1$, p', p_1, ..., $p_n \geq 0$ and hence p' is known even though the event A is not known to the experimenter. If the system of numbers $p_i \geq 0$, $i=1,...,n$, $0 < \sum_{i=1}^{n} p_i \leq 1$ is called a *generalized probability distribution* then the following system of numbers $a \leq p_i \leq b$, $i=1,...,n$, $c \leq \sum_{i=1}^{n} p_i \leq d$, where a,b,c,d are fixed constants, may be called a doubly generalized probability distribution. The constants a,b,c,d can be fixed such that the characterization theorems of the various functions considered in Rényi(1961) become still simpler. Therefore, in subsequent sections of this chapter we will only define the various analogous measures of entropy, directed divergence and inaccuracy. As most of the properties, except a few such as recursivity, are the same for these measures as those of the corresponding quantities defined for probability distributions, we will not mention the properties again in this chapter. Also some of the properties are given in the exercises at the end of this chapter.

(a) *MEASURES ANALOGOUS TO ENTROPIES*: Here we define measures analogous to the measures of entropies of chapter 1. Let $(p_1,...,p_n)$, $p_i \geq 0$, $i=1,...,n$, $\sum_{i=1}^{n} p_i \leq 1$, then a measure analogous to Shannon's entropy is defined as follows.

DEFINITION 5.1.1

(5.1.1) $\underline{H}_n(p_1,...,p_n) = -\Sigma p_i \log_2 p_i / \Sigma p_i$

Also the measures analogous to entropies of order α, which are given in (1.2.1) and (1.2.2), are as follows.

DEFINITION 5.1.2

(5.1.2) $\underline{\hat{H}}_{n,\alpha}(p_1,...,p_n) = (1-\alpha)^{-1} \log_2(\Sigma p_i^\alpha / \Sigma p_i), \quad \alpha \neq 1$

DEFINITION 5.1.3

(5.1.3) $\underline{H}_{n,\alpha}(p_1,...,p_n) = (\Sigma p_i^\alpha / \Sigma p_i - 1)/(2^{1-\alpha} - 1), \quad \alpha \neq 1$

Measures Not Involving Probabilities   107

Another generalization of (5.1.1) is the following.

DEFINITION 5.1.4

(5.1.4) $\quad \underset{\sim}{H}_n(p_1,\ldots,p_n;\beta) = - \Sigma p_i^{\beta+1} \log p_i / \Sigma p_i$

Characterization theorems for (5.1.1) to (5.1.4) are given by Rényi (1961), Daróczy (1963), Aczél (1964), Vajda (1968) and Rathie (1971b, 1971c, 1973).

(b) *MEASURES ANALOGOUS TO DIRECTED DIVERGENCES:* In this section we will give measures analogous to directed divergences of chapter 2. Let $(p_1,\ldots,p_n)$, $p_i \geq 0$, $\Sigma p_i \leq 1$ and $(q_1,\ldots,q_n)$, $q_i \geq 0$, $\Sigma q_i \leq 1$ be two sequences. Then a measure analogous to directed divergence is defined as follows.

DEFINITION 5.1.5

(5.1.5) $\quad \underset{\sim}{I}_n(p_1,\ldots,p_n;q_1,\ldots,q_n) = \Sigma p_i \log_2(p_i/q_i)/\Sigma p_i$

with the usual conventions of chapter 2. For $\alpha \neq 1$ the analogous measures of directed divergence of order α are the following.

DEFINITION 5.1.6

(5.1.6) $\quad \hat{\underset{\sim}{I}}_{n,\alpha}(p_1,\ldots,p_n;q_1,\ldots,q_n) = (\alpha-1)^{-1} \log_2(\Sigma p_i^{\alpha} q_i^{1-\alpha}/\Sigma p_i)$

and

DEFINITION 5.1.7

(5.1.7) $\quad \underset{\sim}{I}_{n,\alpha}(p_1,\ldots,p_n;q_1,\ldots,q_n) = (\Sigma p_i^{\alpha} q_i^{1-\alpha}/\Sigma p_i - 1)/(2^{\alpha-1} - 1)$

with the usual notations of chapter 2. The quantity in (5.1.5) is further generalized to the following quantity.

DEFINITION 5.1.8

(5.1.8) $\quad \underset{\sim}{I}_n(p_1,\ldots,p_n;q_1,\ldots,q_n;\beta) = \Sigma p_i^{\beta+1} \log(p_i/q_i)/\Sigma p_i$

Axiomatic characterizations of quantities (5.1.5) to (5.1.8) are given in Rényi (1961), Rathie (1971c, 1973) and Kaufman and Rathie (1970).

(c) *MEASURES ANALOGOUS TO INACCURACIES:* In this section we will define measures analogous to inaccuracies of chapter 3 and their α-forms. For $(p_1,\ldots,p_n)$, $p_i \geq 0$, $\Sigma p_i \leq 1$ and $(q_1,\ldots,q_n)$, $q_i \geq 0$, $\Sigma q_i \leq 1$, a measure analogous to inaccuracy is defined as follows.

DEFINITION 5.1.9

(5.1.9) $\quad \underset{\sim}{H}_n(p_1,\ldots,p_n;q_1,\ldots,q_n) = - \Sigma p_i \log q_i / \Sigma p_i$

with the usual convention of chapter 3. The two α-forms are the following:

DEFINITION 5.1.10

(5.1.10) $\underline{H}_{n,\alpha}(p_1,\ldots,p_n;q_1,\ldots,q_n) = (\Sigma p_i q_i^{\alpha-1}/\Sigma p_i - 1)/(2^{1-\alpha} - 1)$, $\alpha \neq 1$

and

DEFINITION 5.1.11

(5.1.11) $\underline{\hat{H}}_{n,\alpha}(p_1,\ldots,p_n;q_1,\ldots,q_n) = (1-\alpha)^{-1} \log_2(\Sigma p_i q_i^{\alpha-1}/\Sigma p_i)$, $\alpha \neq 1$

with the usual convention of chapter 3. Another generalization of (5.1.9) is the following.

DEFINITION 5.1.12

(5.1.12) $\underline{H}_n(p_1,\ldots,p_n;q_1,\ldots,q_n;\beta) = -\Sigma p_i^{\beta+1} \log q_i/\Sigma p_i$

Characterization theorems for (5.1.9) to (5.1.12) are given in Rathie (1971c) and Kaufman and Rathie (1970).

5.2 SOME GENERALIZATIONS

In this section we will define further generalizations of the quantities defined in section 5.1.

(a) MEASURES ANALOGOUS TO GENERALIZED ENTROPIES: For $(p_1,\ldots,p_n)$, $p_i \geq 0$, $\Sigma p_i \leq 1$, the measures in definitions *1.3.1*, *1.3.2* and *1.3.3* of section 1.3 may be taken as such for the purpose of some analogous measures to generalized entropies. Axiomatic characterizations of these measures and some of their special cases are given in Rathie (1970, 1971b, 1972, 1973) and Aczél and Daróczy (1963b).

(b) MEASURES ANALOGOUS TO GENERALIZED DIRECTED DIVERGENCES: Several measures defined in section 2.3 may be taken as such here for $(p_1,\ldots,p_n)$, $p_i \geq 0$, $\Sigma p_i \leq 1$ and $(q_1,\ldots,q_n)$, $q_i \geq 0$, $\Sigma q_i \leq 1$ with the usual convention. For various characterizations, see Rathie (1970a, 1971b, 1973).

(c) MEASURES ANALOGOUS TO DIRECTED DIVERGENCES INVOLVING THREE SEQUENCES: For $(p_1,\ldots,p_n)$, $p_i \geq 0$, $\Sigma p_i \leq 1$, $(q_1,\ldots,q_n)$, $q_i \geq 0$, $\Sigma q_i \leq 1$ and $(r_1,\ldots,r_n)$, $r_i \geq 0$, $\Sigma r_i \leq 1$, we define the following measures analogous to directed divergences defined in section 2.4 of chapter 2.

DEFINITION 5.2.1

(5.2.1) $\underline{I}_n(p_1,\ldots,p_n;q_1,\ldots,q_n;r_1,\ldots,r_n) = \Sigma p_i \log_2(q_i/r_i)/\Sigma p_i$

DEFINITION 5.2.2

(5.2.2) $\underline{I}_{n,\alpha}(p_1,\ldots,p_n;q_1,\ldots,q_n;r_1,\ldots,r_n) = [\Sigma p_i q_i^{\alpha-1} r_i^{1-\alpha}/\Sigma p_i - 1]/(2^{\alpha-1} - 1)$

DEFINITION 5.2.3 $\alpha \neq 1$

Measures Useful in Pattern Recognition

(5.2.3) $\hat{I}_{n,\alpha}(p_1,\ldots,p_n:q_1,\ldots,q_n:r_1,\ldots,r_n) = (\alpha-1)^{-1}\log(\Sigma p_i q_i^{\alpha-1} r_i^{1-\alpha}/\Sigma p_i)$,

$\alpha \neq 1$

For properties, some characterization theorems and further generalizations, see Aczél and Nath (1972), Nath (1968,1970) and Rathie (1970a).

5.3 MEASURES USEFUL IN PATTERN RECOGNITION

For two distributions $P \varepsilon S_n$ and $Q \varepsilon S_n$, Toussaint (1973) has discussed an information measure, which was modified by Rathie (1974) and in the modified form, it is defined as follows:

(5.3.1) $K_n^r(P:Q) = 1 - \Sigma_{i=1}^{n}[2^{1/r} p_i q_i/(p_i^r + q_i^r)^{1/r}]$

where $(p_i,q_i) \neq (0,0)$ and r is a positive real number. An interesting special case of (5.3.1) for $r=1$ is given below.

(5.3.2) $K_n(P:Q) = 1 - \Sigma_{i=1}^{n}[2p_i q_i/(p_i+q_i)]$, $(p_i,q_i) \neq (0,0)$

It may be noted that the quantity within brackets in (5.3.2) is the harmonic mean between p_i and q_i. For $n=2$, (5.3.2) reduces to

(5.3.3) $K_2(p,1-p:q,1-q) = 1 - 2pq/(p+q) - 2(1-p)(1-q)/(2-p-q)$

for $p,q \varepsilon [0,1]$ with $(p,q) \neq (0,0)$ or $(1,1)$. The following characterization theorem is due to Rathie (1974).

THEOREM 5.3.1: Let K_n, $n=2,3,\ldots$ satisfy the following postulates.

A1. *Recursivity*: For p_1+q_1, p_2+q_2, p_1+p_2, $q_1+q_2 > 0$ and for all $n \geq 3$,

(5.3.4) $K_n(p_1,\ldots,p_n:q_1,\ldots,q_n) = K_{n-1}(p_1+p_2,p_3,\ldots,p_n:q_1+q_2,q_3,\ldots,q_n)$

$+ [2(2p_1 q_1 + p_1 q_2 + q_1 p_2)(2p_2 q_2 + p_1 q_2 + q_1 p_2)/\{(p_1+p_2+q_1+q_2)(p_1+q_1)$

$(p_2+q_2)\}] K_2(p_1/(p_1+p_2),p_2/(p_1+p_2):q_1/(q_1+q_2),q_2/(q_1+q_2))$;

A2. *Symmetry*: If $\{a_1,a_2,a_3\}$ is any arbitrary permutation of $\{1,2,3\}$, then

(5.3.5) $K_3(p_1,p_2,p_3:q_1,q_2,q_3) = K_3(p_{a_1},p_{a_2},p_{a_3}:q_{a_1},q_{a_2},q_{a_3})$;

A3. *Normalization*

(5.3.6) $K_2(1,0:1/2,1/2) = 1/3$.

Then K_n is uniquely given by (5.3.2).

Proof: Let

(5.3.7) $f(x,y) = K_2(x,1-x:y,1-y)$, $x,y \varepsilon [0,1]$, $(x,y) \neq (0,0)$ or $(1,1)$

Then from postulate A1 for $n=3$ and postulate A2, it is easy to see that

(5.3.8) $f(x,y) = f(1-x,1-y)$, $x,y \in [0,1]$, $(x,y) \neq (0,0), (1,1)$

Clearly (5.3.6), (5.3.7) and (5.3.8) give

(5.3.9) $f(0,1/2) = f(1,1/2) = 1/3$

Again from postulates A2, A1 for $n=3$, (5.3.7) and (5.3.8) one can easily derive the following functional equation.

(5.3.10) $f(x,y) + [2\{u(1-y) + v(1-x)\}\{(1-x-u)(1-y) + (1-y-v)(1-x)\}/\{(2-x-y)$
$(u+v)(2-x-y-u-v)\}] f(u/(1-x),v/(1-y))$
$= f(u,v) + [2\{x(1-v)+y(1-u)\}\{(1-x-u)(1-v)+(1-y-v)(1-u)\}/\{(2-u$
$-v)(x+y)(2-x-y-u-v)\}] f(x/(1-u),y/(1-v))$

for $x,y,u,v \in [0,1[$ with $x+u, y+v \in [0,1]$ such that $(x,y) \neq (0,0)$ and $(u,v) \neq (0,0)$.

Following the method of Rathie and Kannappan (1972), the solution of the equation (5.3.10) under the condition (5.3.9) can be derived as

(5.3.11) $f(x,y) = 1 - 2xy/(x+y) - 2(1-x)(1-y)/(2-x-y)$

for $x,y \in [0,1]$ with $(x,y) \neq (0,0)$ or $(1,1)$. The successive applications of the postulate A1 gives

(5.3.12) $K_n(P:Q) = \sum_{i=2}^{n} [2(r_{i-1} s_i + s_{i-1} r_i)(p_i s_i + q_i r_i)/\{(r_i + s_i)(r_{i-1}$
$+ s_{i-1})(p_i + q_i)\}] K_2(p_i/r_i, r_{i-1}/r_i : q_i/s_i, s_{i-1}/s_i)$

for $(p_i,q_i) \neq (0,0)$ where $r_i = p_1+\ldots+p_i$ and $s_i = q_1+\ldots+q_i$ for all $i = 2,3,\ldots,n$. Now (5.3.12) with the help of (5.3.7) and (5.3.11) and after a little simplification. proves the theorem.

5.4   INEQUALITIES AMONG INFORMATION MEASURES

In this section we discuss the various inequalities among information measures defined for continuous cases. These results are true for discrete cases as well. In order to make this section self contained a few of the information measures are also defined here.

Let $p(x)$ and $q(x)$ be two probability density functions which could characterize two pattern classes. Then we define the following measures of information. In what follows, logarithmns are taken to the base e.

DEFINITION 5.4.1: *Discrimination Information, Directed Divergence or Information Gain.*

Inequalities Among Information Measures

(5.4.1) $\quad I = I_{12} = \int p(x) \log(p(x)/q(x)) dx$

(5.4.2) $\quad I_{21} = \int q(x) \log(q(x)/p(x)) dx$

DEFINITION 5.4.2: *Divergence*

(5.4.3) $\quad J = \int \{p(x) - q(x)\} \log\{p(x)/q(x)\} dx$

In the case of multiclass problem,

(5.4.4) $\quad \bar{J} = (1/M^2) \sum_{i=1}^{M} \sum_{j=1}^{M} \int \{p_i(x) - p_j(x)\} \log\{p_i(x)/p_j(x)\} dx$

DEFINITION 5.4.3: *Kolmogorov's Variational Distance*

(5.4.5) $\quad V = \int |p(x) - q(x)| \, dx$

Some generalizations of (5.4.4) are given in chapter 4.

DEFINITION 5.4.4: *Bhattacharyya Coefficient of Affinity*

(5.4.6) $\quad \rho = \int \{p(x) q(x)\}^{\frac{1}{2}} dx$

(5.4.7) $\quad \rho_M = \int \{p_1(x) p_2(x) \ldots p_M(x)\}^{1/M} dx$

These measures and their generalizations were mentioned in chapters 2 and 4.

DEFINITION 5.4.5: *Harmonic-Mean Coefficient*

(5.4.8) $\quad T_1 = \int 2p(x)q(x)/\{p(x) + q(x)\} \, dx$

Now we will discuss, without proofs, various inequalities among information measures.

(a) INEQUALITIES BETWEEN I_{12} AND V

Volkonskij and Rozanov (1959) proved that

(5.4.9) $\quad I \geq V - \log(1+V)$

Pinsker (1960) improved (5.4.9) by establishing that

(5.4.10) $\quad I \geq V^2/a$

where $a > 2$ is a constant. Csiszár (1966) proved (5.4.10) with $a=16$ while McKean (1966) derived (5.4.10) with $a=4e$. Csiszár (1967) and Kemperman (1969) sharpened these results by proving

(5.4.11) $\quad I \geq V^2/2$

Kullback (1967, 1970) improved (5.4.11) by showing that

(5.4.12) $\quad I \geq V^2/2 + V^4/36$

For V close to 2 the bounds given in (5.4.9), (5.4.10), (5.4.11) and (5.4.12) are loose and for $V=2$ the equality does not hold. Vajda

(1970) improved the above bounds at least for V close to 2 and showed that

(5.4.13) $I \geq log\{(2+V)/(2-V)\} - 2V(2+V)$

The bound given in (5.4.13) is slightly weaker than (5.4.12) for V less than 1.75 approximately and much sharper than (5.4.12) for V greater than 1.75 approximately. Also equality holds in (5.4.13) for $V=0$ and $V=2$.

Toussaint (1974) sharpened (5.4.12) by proving the following result.

(5.4.14) $I \geq V^2/2 + V^4/36 + V^6/288$

Thus, the maximum of (5.4.13) and (5.4.14) provide the sharpest lower bound available for I in terms of V for arbitrary distributions. For the case of Gaussian distributions with equal covariance matrices, Toussaint (1974) showed that

(5.4.15) $I \geq (V/2)log\{(2+V)/(2-V)\}$

where the equality holds for both $V=0$ and $V=2$. Furthermore, it was shown that (5.4.15) is sharper than both (5.4.13) and (5.4.14) for $V \in [0,2]$.

(b) INEQUALITIES BETWEEN J AND V

From (5.4.15) it follows that

(5.4.16) $J \geq V\, log\{(2+V)/(2-V)\}$

(c) INEQUALITIES BETWEEN J AND ρ

Ikeda (1963) showed that

(5.4.17) $J \geq 2(1 - \rho^2)^{\frac{1}{2}}$

which was improved by Toussaint (1972) as

(5.4.18) $J \geq 4(1 - \rho)$

Hoeffding and Wolfowitz (1958) proved the result

(5.4.19) $J \geq - 4\, log\, \rho$

which was generalized by Toussaint (1974a) to show that

(5.4.20) $J(\pi_1, \pi_2) \geq - 2\{H(\pi) + log\, \rho(\pi_1, \pi_2)\}$

where π_i, $i=1,2$ are the a priori probabilities of occurrence of an event distributed according to $p(x)$ and $q(x)$, $\pi_1 + \pi_2 = 1$;

(5.4.21) $J(\pi_1, \pi_2) = \int \{\pi_1 p(x) - \pi_2 q(x)\} log\{\pi_1 p(x)/\pi_2 q(x)\} dx$

(5.4.22) $H(\pi) = -\pi_1 \log \pi_1 - \pi_2 \log \pi_2$

and

(5.4.23) $\rho(\pi_1,\pi_2) = \int (\pi_1 p(x) \pi_2 q(x))^{\frac{1}{2}} dx$

(d) INEQUALITIES BETWEEN J AND T_1

Toussaint (1974a) has proved the following bounds.

(5.4.24) $T_1(\pi_1,\pi_2) \geq 2 \exp\{-2H(\pi) - J(\pi_1,\pi_2)\}$

where

(5.4.25) $T_1(\pi_1, \pi_2) = \int \{2\pi_1 p(x)/(\pi_2 q(x))\}/\{\pi_1 p(x) + \pi_2 q(x)\} dx$

For $\pi_1 = \pi_2 = 1/2$, (5.4.24) reduces to

(5.4.26) $T_1 \geq e^{-J/2}$

where the equality holds for both $J=0$ and $J=\infty$.

(5.4.27) $T_1(\pi_1,\pi_2) \geq (1/2)\{1 - (1/2)J(\pi_1,\pi_2)\}$

For $\pi_1 = \pi_2 = 1/2$, (5.4.27) reduces to

(5.4.28) $T_1 \geq 1 - J/4$

where the equality holds when $J=0$.

An inequality that is sharper than (5.4.24) and (5.4.27) is given by

(5.4.29) $J(\pi_1,\pi_2) \geq (1 - 2T_1(\pi_1,\pi_2))^{\frac{1}{2}} \log[\{1 + (1-2T_1(\pi_1,\pi_2))^{\frac{1}{2}}\}/\{1-(1-$

$-2T_1(\pi_1,\pi_2))^{\frac{1}{2}}\}]$

which for $\pi_1 = \pi_2 = 1/2$ reduces to

(5.4.30) $J \geq 2(1-T_1)^{\frac{1}{2}} \log[\{1+(1-T_1)^{\frac{1}{2}}\}/\{1-(1-T_1)^{\frac{1}{2}}\}]$

(e) INEQUALITIES BETWEEN V AND ρ

Matusita (1955) showed that

(5.4.31) $2(1-\rho) \leq V \leq 2\{2(1-\rho)\}^{\frac{1}{2}}$

which was improved by Matusita (1956) by showing that

(5.4.32) $V \leq 2(1-\rho^2)^{\frac{1}{2}} \leq 2\{2(1-\rho)\}^{\frac{1}{2}}$

Kailath (1967) generalized (5.4.32) to show that

(5.4.33) $1 - 2\rho(\pi_1,\pi_2) \leq V(\pi_1,\pi_2) \leq \{1 - 4(\rho(\pi_1,\pi_2))^2\}^{\frac{1}{2}}$

where

(5.4.34) $V(\pi_1,\pi_2) = \int |\pi_1 p(x) - \pi_2 q(x)| dx$

(f) INEQUALITY BETWEEN V AND T_1

Toussaint (1974a) showed that an inequality between V and T_1 follows directly from a result of Cover and Hart (1967). In particualr,

(5.4.35) $1 - V/2 \leq T_1 \leq 1 - V^2/4$

(g) INEQUALITIES BETWEEN ρ AND T_1

Horibe (1970) proved a lower bound on T_1. He showed that

(5.4.36) $T_1(\pi_1,\pi_2) \geq 2\{\rho(\pi_1,\pi_2)\}^2$

which for $\pi_1=\pi_2=1/2$ reduces to

(5.4.37) $T_1 \geq \rho^2$

Ito (1972) proved an upper bound on T_1, namely

(5.4.38) $T_1(\pi_1,\pi_2) \leq \rho(\pi_1,\pi_2)$

which for $\pi_1=\pi_2=1/2$ reduces to

(5.4.39) $T_1 \leq \rho$

(h) INEQUALITY BETWEEN $\bar{J}$ AND ρ_M

Toussaint (1974b, 1974c) showed that

(5.4.40) $\bar{J} \geq -2 \log \rho_M$

For $M=2$, (5.4.40) reduces to (5.4.19).

5.5 APPLICATIONS

Consider the problem of pattern recognition. Let X be a measurement vector resulting from some pattern P; P is classified as belonging to class i if

$$\pi_i p_i(x) > \pi_j p_j(x), \quad i,j = 1,2, \quad i \neq j$$

where the π's denote prior probabilities. This is the usual optimal Bayes' rule which yields the probability of misclassification given by

$$p_e = \int \min_i \{\pi_i p_i(x)\} \, dx, \quad i = 1,2$$

Bounds on p_e, in terms of the various information measures, may be seen from Hellman and Raviv (1970), Toussaint (1974a,1974c, 1974d), Kailath (1967), Matusita (1967), Cover and Hart (1967) and Hudimoto (1956). Applications of some measures of information which arise in the context of statistical games can be seen from Gottinger (1974).

5.6 OPEN PROBLEMS

5.1 Solve the following functional equation

$$f(x,y) + g(x,y,u,v)f(u/(1-x),v/(1-y))$$
$$= f(u,v) + h(x,y,u,v)f(x/(1-u),y/(1-v))$$

for $x,y,u,v \in [0,1[$, $x+u$, $y+v \in [0,1]$ where g and h are given functions.

5.2 Extend 5.1 to k variables $x_1, x_2, \ldots, x_k$. Discuss the case when $g = h$.

EXERCISES

5.1 (Rényi,1961, Daróczy,1963). Let Δ be the set of all sequences $P=(p_1,\ldots,p_n)$ of non-negative numbers such that $0 < \Sigma p_i \leq 1$. Let
- A1. $H(P)$ be a symmetric function of the elements of P;
- A2. $H(P)$ be a continuous function of $p \in]0,1]$;
- A3. $H(½) = 1$;
- A4. For $P \in \Delta$ and $Q \in \Delta$, let $H(P*Q) = H(P) + H(Q)$, where $P*Q$ consists of the numbers $p_i q_j$, $i=1,\ldots,n$, $j=1,\ldots,m$;
- A5. Let there exist a strictly monotonic and continuous function $g(x)$ such that if $P \in \Delta$, $Q \in \Delta$ and $W(P) + W(Q) \leq 1$, let

$$H(P \cup Q) = g^{-1}\left[\frac{W(P)g\{H(P)\} + W(Q)g\{H(Q)\}}{W(P) + W(Q)}\right],$$

where $P \cup Q = (p_1,\ldots,p_n, q_1,\ldots,q_m)$, $W(P) = \Sigma_{i=1}^n p_i$ and $W(Q) = \Sigma_{j=1}^m q_j$. Then show that $g(x)$ is either linear or exponential and that (5.1.1) or (5.1.2) is then uniquely determined.

5.2 (Aczél 1964). For $(p_1,\ldots,p_n)$, $p_i > 0$, $i=1,\ldots,n$, $\Sigma_{i=1}^n p_i \leq 1$ let $H(p_1,\ldots,p_n)$ satisfy the following postulates.
- B1. $H(P)$ is continuous in $]0,1[$;
- B2. $H(½) = 1$;
- B3. $H(p_1 q_1,\ldots,p_1 q_m,\ldots,p_n q_1,\ldots,p_n q_m) = H(p_1,\ldots,p_n) + H(q_1,\ldots,q_m)$, for $p_i, q_j > 0$, $\Sigma_{i=1}^n p_i \leq 1$, $\Sigma_{j=1}^m q_j \leq 1$;
- B4. $H(p_1,\ldots,p_n,q_1,\ldots,q_m) \ g^{-1}([\Sigma_{i=1}^n p_i \ g\{H(p_1,\ldots,p_n)\} + \Sigma_{j=1}^m q_j \ g\{H(q_1,\ldots,q_m)\}]/[\Sigma_{i=1}^n p_i + \Sigma_{j=1}^m q_j])$. for $\Sigma_{i=1}^n p_i + \Sigma_{j=1}^m q_j \leq 1$, $p_i > 0$, $q_j > 0$, $i=1,\ldots,n$, $j=1,\ldots,m$, where g is continuous and strictly monotonic. Then show that either $H(p_1,\ldots,p_n)$ is given by (5.1.1) or (5.1.2).

5.3 (Vajda, 1968). Let Δ denote all the sets $(p_1,\ldots,p_n)$, $p_i \geq 0, \Sigma p_i \leq 1$, $i = 1,2,\ldots,n$. Prove that the following axioms uniquely determine (5.1.3) for $\alpha \neq 1$.

116   Some Other Measures and Inequalities

C1. $\lim_{p \to 0} \underline{H}(1-p)/p = constant;$

C2. $\underline{H}(½) = 1;$

C3. $\underline{H}(P)$ is a symmetric function of all its parameters, $p_1, p_2, \ldots, p_n$;

C4. For every $P=(p_1,\ldots,p_n)$, $Q=(q_1,\ldots,q_m) \in \Delta$, $\underline{H}(P*Q) = \underline{H}(P) + \underline{H}(Q) + (2^{1-\alpha}-1)\underline{H}(P)\underline{H}(Q)$;

C5. For $P,Q \in \Delta$, $W(P) + W(Q) \leq 1$, $\underline{H}(P \cup Q) = \{W(P)\underline{H}(P) + W(Q)\underline{H}(Q)\}/\{W(P)+W(Q)\}$.

5.4 (Kaufman and Rathie, 1970). Show that the functions $\underline{H}(p_1,\ldots,p_n:q_1,\ldots,q_n)$ for $p_i > 0$, $\Sigma p_i \leq 1$, $q_i > 0$, $\Sigma q_i \leq 1$ satisfying the following postulates have exactly two forms given by (5.1.4) or (5.1.10).

D1. $\underline{H}(p:1)$ and $\underline{H}(1:q)$ are continuous functions of p and q respectively where $p,q \in]0,1]$;

D2. $\underline{H}(1:½) = 1;$

D3. $\underline{H}(½:1) = 0;$

D4. If $P = P_1*P_2$ and $Q = Q_1*Q_2$, then $\underline{H}(P:Q) = \underline{H}(P_1:Q_1)+\underline{H}(P_2:Q_2)$, where the correspondence between the elements of P and Q is that induced by the correspondence between the elements of P_1 and Q_1 and those of P_2 and Q_2 ;

D5. There exists a continuous and strictly monotonic increasing function $g(x)$ defined for all real x such that $H(P:Q) = g^{-1}([W(P_1)g\{H(P_1:Q_1)\} + W(P_2)g\{H(P_2:Q_2)\}]/[W(P_1) + W(P_2)])$ where $P = P_1 \cup P_2$ and $Q = Q_1 \cup Q_2$.

5.5 (Rényi, 1961). For $P = (p_1,\ldots,p_n)$, $p_i > 0$, $\Sigma p_i \leq 1$, $Q = (q_1,\ldots,q_n)$, $q_i > 0$, $\Sigma q_i \leq 1$, let $\underline{I}(P:Q)$ satisfy the following postulates.

E1. $\underline{I}(P:Q)$ is unchanged if the elements of P and Q are rearranged in the same way so that the one-to-one correspondence between them is not altered;

E2. If $p_i \leq q_i$, $i=1,\ldots,n$ then $\underline{I}(P:Q) \leq 0$ while if $p_i \geq q_i$, then $\underline{I}(P:Q) \geq 0$;

E3. $\underline{I}(½:1) = 1;$

E4. Same as D4 with $\underline{H}$ replaced by $\underline{I}$;

E5. Same as D5 with $\underline{H}$ replaced by $\underline{I}$.

Then show that the function g is either a linear or an exponential function. In the first case prove that $\underline{I}$ is given by (5.1.5) while in the second case by (5.1.6).

5.6 (Aczél and Nath, 1972). Let $I(p:q:r)$ and $I(p_1,p_2:q_1,q_2:r_1,r_2)$ be defined for $p,p_1,p_2,q,q_1,q_2,r,r_1,r_2, p_1+p_2, q_1+q_2, r_1+r_2 \in]0,1]$.

Exercises

Consider the following postulates.

F1. $I(p:q:r) >, =, < 0$ according as $q >, =, < r$, $p,q,r \in]0,1]$;

F2. $I(1:1:\frac{1}{2}) = 1$;

F3. $I(p_1x, p_2x,\ldots,p_nx:q_1y,\ldots,q_ny:r_1z,\ldots,r_nz) = I(p_1,\ldots,p_n:q_1,\ldots,q_n:r_1,\ldots,r_n) + I(x:y:z)$, for $n = 1,2$;

F4. If p_1,p_2,q_1,q_2,r_1,r_2 are positive such that $p_1+p_2 \leq 1$, $q_1+q_2 \leq 1$, $r_1+r_2 \leq 1$, then there exists a continuous and strictly monotonic real function ϕ such that $I(p_1,p_2:q_1,q_2:r_1,r_2) = \phi^{-1}[(p_1\phi\{I(p_1:q_1:r_1)\} + p_2\phi\{I(p_2:q_2:r_2)\})/(p_1+p_2)]$.

Then show that F1 to F4 uniquely determine the function $I(p:q:r)$ as $I(p:q:r) = \log_2(q/r)$ and ϕ in postulate F4 is either a linear function or a linear function of an exponential function. In the first case, $I=\underset{\sim}{I}_2$ given by (5.2.1) and in the second case $I=\underset{\sim}{I}_{2,\alpha}$ given by (5.2.2).

5.7 (Arlotti, 1970-71). For every $P=(p_1,\ldots,p_n)$, $p_i \geq 0$, $0 < \Sigma p_i \leq 1$, let the function $K_n(p_1,\ldots,p_n)$ satisfy the following postulates.

G1. $K(p)$ is continuous for $p \in]0,1]$ and non-negative;

G2. $K_n(p_1,\ldots,p_n)$ is a symmetric function of its variables;

G3. $K_{n+1}(p_1,\ldots,p_n,0) = K_n(p_1,\ldots,p_n)$;

G4. $K_n(p_1,\ldots,p_n) - K_{n-1}(p_1,\ldots,p_{n-2},p_{n-1}+p_n) = \phi(p_1+\ldots+p_{n-2},p_{n-1},p_n)$ with $\phi \geq 0$;

G5. $K_{mn}(p_1q_1,\ldots,p_nq_1,\ldots,p_1q_m,\ldots,p_nq_m) = K_n(p_1,\ldots,p_n) + K_m(q_1,\ldots,q_m)$;

G6. $\phi(0,p,1-p)$ is a Lebesgue integrable function over the interval $0 \leq p \leq 1$ not identically equal to zero.

Then prove that for some $a>0$ and $n>1$, K_n is given by

$$K_n(p_1,\ldots,p_n) = -\Sigma_{i=1}^n p_i \log_a p_i/\Sigma_{i=1}^n p_i + \log_a(\Sigma_{i=1}^n p_i) - \log_b(\Sigma_{i=1}^n p_i)$$

REFERENCES

Aczél, J. (1964). Zur gemeinsamen charakterisierung der entropien α-ter ordnung und der Shannonschen entropie nicht unbedingt vollständiger verteilungen. *Z. Wahr. Verw. Geb.*, 3, 177-183.

Aczél, J. (1966). *Lectures on Functional Equations and Their Applications*. Academic Press, New York.

Aczél, J. (1968). On different characterizations of entropies. *Proc. Internat. Symp. Probability and Information Theory*. McMaster University. *Lecture Notes in Mathematics*, Vol. 89, Springer-Verlag, 1969, 1-11.

Aczél, J., Eaker, J.A., Djoković, D.Ž., Kannappan, Pl. and Rado, F. (1971). Extensions of certain homomorphisms of sub-semigroups to homomorphisms. *Aeq. Math.*, 6, 263-271.

Aczél, J. and Daróczy, Z. (1963). Charakterisierung der entropien positiver ordnung und der Shannonschen entropie. *Acta Math. Acad. Sci. Hungar.*, 14, 95-121.

Aczél, J. and Daróczy, Z. (1963a). Sur la caractérisation axiomatique des entropies d'ordre positif, y comprise l'entropie de Shannon. *C.R. Acad. Sci. Paris*, 257, 1581-1584.

Aczél, J. and Daróczy, Z. (1963b). Über verallgemeinerte quasilineare mittelwerte, die mit dewichtsfunktionen gebildet sind. *Publicationes Mathematicae*, 10, 171-190.

Aczél, J., Forte, B. and Ng, C.T. (1974). Why the Shannon and Hartley entropies are "natural"? *Adv. Appl. Prob.*, 6, 131-146.

Aczél, J. and Nath, P. (1972). Axiomatic characterizations of some measures of divergence in Information. *Z. Wahr. Verw. Geb.*, 21, 215-224.

Aczél, J. and Pfanzagl, J. (1966). Remarks on the measurement of subjective probability and information. *Metrika*, 11, 91-105.

Adke, S.R. (1958). A note on distance between two populations. *Sankhya*, 19, 195-200.

Ali, S.M. and Silvey, S.D. (1966). A general class of coefficients of divergence of one distribution from another. *J. Roy. Statist. Soc. Ser. B*, 28, 131-142.

Arlotti, L. (1970-71). On a characterization of the Rényi-Shannon entropy for incomplete probability distributions. *Publ. Math. Debrecen*, 17, 35-40.

Ash, R. (1965). *Information Theory*. Wiley, New York.

Attneave, F. (1959). *Applications of Information Theory to Psychology*. Holt, Rinehart and Winston, New York.

Baiocchi, C. and Pintacuda, N. (1968). Sul l'assiomatica della teoria del l'informazione. *Annali de Matematica Pura ed Applicata*, 80, 301-326.

Balasubrahmanyam, P. and Siromoney, G. (1968). A note on entropy of Telungu prose. *Information and Control*, 13, 281-285.

Beckenbach, E.F. and Bellman, R. (1965). *Inequalities*. Springer-Verlag, New York.

Belis, M. and Guisasu, S. (1968). Quantitative-qualitative measure of information in cybernetic systems. *IEEE Transactions on Information Theory*, 591-592.

Bendig, A.W. (1953). Twenty questions: an informational analysis. *J. Exp. Psychology*, 46, 345-348.

Bhargava, T.N. and Doyle, P.H. (1974). A geometric study of diversity. *J. Theor. Biol.*, 43, 241-251.

Bhargava, T.N. and Uppuluri, V.R.R. (1971). On diversity in human ecology. *Research Report*, Kent State University.

Bhattacharyya, A. (1945-46). On a measure of divergence between two multinomial populations. *Sankhya*, 7, 401-406.

Blachman, N.M. (1967). The amount of information y gives about x. *Unpublished Report*.

Borges, R. (1967). Zur Herleitung der Shannonschen Information. *Math. Z.*, 96, 282-287.

Brillouin, L. (1956). *Science and Information Theory*. Academic Press, New York.

Campbell, L.L. (1965). A coding theorem and Rényi's entropy. *Information and Control*, 8, 423-429.

Campbell, L.L. (1966). Definition of entropy by means of a coding problem. *Z. Wahr. Verw. Geb.*, 6, 113-118.

References

Campbell, L.L.(1970). Equivalence of Gauss' principle and minimum discrimination information estimation of probabilities. *Ann. Math. Statist.*, 41, 1011-1015.

Campbell, L.L.(1972). Characterization of entropy of probability distribution on the real line. *Information and Control*, 21, 329-338.

Campbell, L.L.(1974). Information divergence:probabilistic properties and axiomatic characterizations. *Symp. Measures of Information and Their Applications*. Indian Institute of Technology Bombay, August 16-18,1974.

Chaundy, T.W. and McLeod, J.B.(1960). On a functional equation. *Proc. Edinburgh Math. Soc. Notes*, 43, 7-8.

Cover, T.M. and Hart, P.E.(1967). Nearest neighbour pattern classification. *IEEE Trans. Information Theory*, IT-13, 21-27.

Cozzolino, J.M. and Zahner, M.J.(1973). The maximum entropy distribution of the future market price of a stock. *Operations Res.*, 21 No.6, 1200-1211.

Csiszár, I.(1961). Some remarks on the dimension and entropy of random variables. *Acta Math. Acad. Sci. Hungar*, 12, 399-408.

Csiszár, I.(1962). On the dimension and entropy of order α of the mixture of probability distributions. *Acta Math. Acad. Sci. Hungar.*, 13, 245-255.

Csiszár, I.(1966). A note on Jensen's inequality. *Studia Sci.Math.Hungar.*, 1, 185-188.

Csiszár, I.(1967). Information-type measures of difference of probability distributions and indirect observations. *Studia Sci. Math. Hungar.*, 2, 299-318.

Csiszár, I.(1967a). On topological properties of f-divergences. *Studia Sci. Math. Hungar.*, 2, 329-337.

Daróczy, Z.(1963). Über die gemeinsame charakterisierung der zu den nicht vollständigen verteilungen gehörigen entropien von Shannon und Rényi. *Z. Wahr. Verw. Geb.*, 1, 381-388.

Daróczy, Z.(1964). Über mittelwerte und entropien vollständiger wahrscheinlichkeitsverteilung. *Acta Math. Acad. Sci. Hungar.*, 15, 203-210.

Daróczy, Z.(1967). Über eine charakterisierung der Shannonschen entropie. *Statistica*, 27, 199-205.

Daróczy, Z.(1967a). Über die charakterisierung der Shannonschen entropie. *Proc. Coll. Information Theory.(September 19-24, 1967)*, 135-139.

Daróczy, Z.(1969). Über ein funktionalgleichungssystem der informationstheorie. *Aeq. Math.*, 2, 144-149.

Daróczy, Z.(1969a). On the Shannon measure of information (Hungarian), *Magyar Tud. Akad. Mat. Fiz. Oszt. Közl.*, 19, 9-24.

Daróczy, Z.(1970). Generalized information functions. *Information and Control*, 16, 36-51.

Daróczy, Z.(1971). On the measurable solution of a functional equation. *Acta Math. Acad. Sci. Hungar.*, 22, 11-18.

Daróczy, Z and Katai, I.(1970). Additive zahlentheoretische funktionen und das mass der information. *Ann. Univ. Sci. Budapest. Eötvös Sect. Math.*, 13, 83-88.

Dutta, M.(1966). On maximum (information-theoretic) entropy estimation. *Sankhya*, 28, 319-328.

Erdös, P.(1958). On the distribution of additive arithmetical functions. *Rend. Sem. Mat. Fis. Milano*, 27, 3-7.

Faddeev, D.K.(1956). On the concept of entropy of a finite probabilistic scheme (Russian). *Uspeki Mat. Nauk.*, 11, No.1, 227-231.

Fano, R.M.(1949). The transmission of information. *M.I.T. R.L.E. Technical Report*, 65.

Feinstein, A.(1958). *Foundations of Information Theory*. McGraw-Hill, New York.

Fischer, P.(1972). On the inequality $\Sigma p_i f(p_i) \geq \Sigma p_i f(q_i)$. *Metrika*, 18, 199-208.

Forte, B.(1973). Why Shannon's entropy? *Convegno Inform. Teor., Ist. Naz. Alta. Mat., Roma 1973. Symposia Math. Vol.XI*, Academic Press, New York, 1974.

Forte, B. and Daróczy, Z.(1968). A characterization of Shannon's entropy. *Boll. U.M.I.*, 4, 631-635.

Forte, B. and Ng, C.T.(1973). On a characterization of the entropies of degree β. *Utilitas Mathematica*, 4, 193-205.

Forte, B. and Pintacuda, N.(1968). Sull informazione associate alle

References

esperience incomplete. *Ann. Mat. Pura. Appl.*, 80(4), 215-224.

Gallager, R.G. (1968). *Information Theory and Reliable Communication*. Wiley, New York.

George, A. and Mathai, A.M. (1974). Applications of the concepts of affinity and distance to population problems. *J. Biosocial Sciences*, 6, 347-356.

Gini, C. (1912). Variabilita e mutabilita. *Studi Economico-Giuridici della Facolta di Giurisprudenza dell Universita di Cagliari*, a III, Parte II.

Gini, C. and Zappa, G. (1937-39). Sulle proprieta delle medie potenziate e combinatorie. *Metron*, 13, 21-31.

Good, I.J. (1966). A derivation of probabilistic explication of information. *J. Roy. Statist. Soc. Ser.B*, 28, 578-581.

Gottinger, H.W. (1974). Some measures of information arising in statistical games. *Kybernetika*, 15, 111-116.

Hart, P.E. (1971). Entropy and other measures of concentration. *J. Roy. Statist. Soc. Ser.A*, 134, 73-85.

Hartley, R.V.L. (1928). Transmission of information. *Bell System Tech. J.*, 7, 535-563.

Havrda, J. and Charvát, F. (1967). Quantification method of classification processes: Concept of structural α-entropy. *Kybernetika*, 3, 30-35.

Hellman, M.E. and Raviv, J. (1970). Probability of error equivocation and the Chernoff bound. *IEEE Trans. Information Theory*, IT-16, 368-372.

Hewitt, E. and Stromberg, K. (1965). *Real and Abstract Analysis*. Springer-Verlag, Berlin and New York.

Hobson, A. (1969). A new theorem of information theory. *J.Statist.Phy.*, 1, 383-391.

Hoeffding, W. and Wolfowitz, J. (1958). Distinguishability of sets of distributions. *Ann. Math. Statist.*, 29, 700-718.

Horibe, Y. (1970). On zero error probability of binary decisions. *IEEE Trans. Information Theory*, IT-16, 347-348.

Horowitz, I. (1970). Employment concentration in the common market: An entropy approach. *J. Roy. Statist. Soc. Ser. A*, 133, 463-479.

Hudimoto, H. (1956). On the distribution-free classification of an individual into one of two groups. *Ann. Inst. Statist. Math.*, 8, 105-112.

Hyvårinen, L.P. (1970). *Information Theory for Systems Engineers*. Springer-Verlag, New York.

Ikeda, S. (1963). Asymptotic equivalence of probability distributions with application to some problems of asymptotic independence. *Ann. Inst. Statist. Math.*, 15, 87-116.

Ito, T. (1972). Approximate error bounds in pattern recognition. *Machine Intelligence (Editors: B.Meltzer and D.Michie)* Edinburgh University Press, 369-376.

Jaynes, E.T. (1957). Information theory and statistical mechanics. *Phys. Rev.*, 106, 620-630.

Jeffreys, H. (1961). *Theory of Probability*. 3rd Ed. Oxford at the Clarenden Press.

Jelinek, F. (1968). Buffer overflow in variable length coding of fixed rate sources. *IEEE Trans. Information Theory*, IT-14, 490-501.

Kailath, T. (1967). The divergence and Battacharyya distance measures in signal selection. *IEEE Trans. Communication Technology*, COM-15, 52-60.

Kannappan, Pl. (1972). On Shannon's entropy, directed-divergence and inaccuracy. *Z. Wahr. Verw. Geb.*, 22, 95-100.

Kannappan, Pl. (1972a). On directed-divergence and inaccuracy. *Z.Wahr. Verw. Geb.*, 25, 49-55.

Kannappan, Pl. (1973). On generalized directed-divergence. *Funkcialaj Ekvacioj*, 16, 71-77.

Kannappan, Pl. and Ng, C.T. (1973). Measurable solutions of functional equations related to information theory. *Proc. Amer. math. Soc.*, 38, 303-310.

Kannappan, Pl. and Ng, C.T. (1974). A functional equation and its application to information theory. *Unpublished Report*.

Kannappan, Pl. and Rathie, P.N. (1971). On the solution of a functional equation connected with inaccuracy. *Communicated for publication*.

Kannappan, Pl. and Rathie, P.N. (1972). An application of a functional equation to information theory. *Ann. Polon.Math.*, 26, 95-101.

Kannappan, Pl. and Rathie, P.N.(1973). On various characterizations of directed-divergence. *Trans. 6th Prague Conference on Information Theory, Statistical Decision Functions, Random Processes. September 19-25,1971.* 331-339.

Kannappan, Pl. and Rathie, P.N.(1973a). On a characterization of directed-divergence. *Information and Control,* 22, 163-171.

Kannappan, Pl. and Rathie, P.N.(1973b). An axiomatic characterization of generalized directed-divergence. *Kybernetika,* 9, 330-337.

Kannappan, Pl. and Rathie, P.N.(1973c). On generalized information functions. *Tôhoku Math. J., (Accepted for publication).*

Kannappan, PL. and Rathie, P.N.(1974). On a generalized directed-divergence function. *Czech. Math. J.,* 24, 5-14.

Kannappan, Pl. and Rathie, P.N.(1974a). On a generalized directed-divergence and related measures. *(Communicated for publication).*

Kapur, J.N.(1967). Generalized entropy of order α and type β. *The mathematical Seminar,* 4, 78-94.

Kapur, J.N.(1967a). Some properties of generalized entropies. *Indian J. Math.,* 9, 427-442.

Kapur, J.N.(1968). Information of order α and type β. *Proc. Indian Acad. Sci.,* A68, 65-75.

Kapur, J.N.(1969). Some properties of entropy of order α and type β. *Proc. Indian Acad. Sci.,* A69, 201-211.

Kapur, J.N.(1972). Measures of uncertainty, mathematical programming and Physics. *J. Ind. Soc. Agri. Res. Statist.,* 24, 47-66.

Kátai, I.(1967). A remark on additive arithmetic functions. *Ann. Univ. Sci. Budapest Eötvös. Sect. Math.,* 12, 81-83.

Kaufman, H. and Mathai, A.M.(1973). An axiomatic foundation for a measure of affinity among a number of distributions. *J. Multivariate Analysis,* 3, 236-242.

Kaufman, H., Mathai, A.M. and Rathie, P.N.(1972). A mathematical foundation for Pearson's X^2 goodness-of-fit statistic. *Sankhya Ser.A.,* 34, 441-442.

Kaufman, H. and Rathie, P.N.(1970). Axiomatic characterization of the measures of inaccuracy and information. *Coll. Math.* (*In Press*).

Kaufman, H. and Rathie, P.N.(1974). Measurable solution of a functional equation concerning inaccuracy. *Symp. Measures of Information and Their Applications*. Indian Institute of Technology, Bombay, August 16-18, 1974.

Kemperman, J.H.B.(1969). On the optimum rate of transmitting information. *Ann. Math. Statist.*, 40, 2156-2177.

Kendall, D.G.(1964). Functional equations in information theory. *Z. Wahr. Verw. Geb.*, 2, 225-229.

Kerridge, D.F.(1961). Inaccuracy and inference. *J. Roy. Statist. Soc. Ser.B*, 23, 184-194.

Khinchin, A.J.(1953). The concept of entropy in the theory of probability.(Russian). *Uspeki. Mat. Nauk.*, 8, No.3(55), 3-20.

Kirmani, S.N.U.A.(1968). Some results on Matusita's measure of distance. *J. Indian Statist. Assoc.*, 6, 89-98.

Kullback, S.(1959). *Information Theory and Statistics*. Wiley, New York.

Kullback, S.(1967). A lower bound for discrimination information in terms of variation. *IEEE Trans. Information Theory*, IT-13, 126-127.

Kullback, S.(1970). Correction to a lower bound for discrimination information in terms of variation. *IEEE Trans. Information Theory*, IT-16, 652.

Kullback, S. and Leibler, R.A.(1951). On information and sufficiency. *Ann. Math. Statist.*, 22, 79-86.

Lee, P.M.(1964). On the axioms of information theory. *Ann. Math. Statist.*, 35, 415-418.

Mallows, C.L.(1959). The information in an experiment. *J. Roy. Statist. Soc. Ser. B*, 21, 67-72.

Mathai, A.M.(1967). Dispersion theory. *Estadistica*, 95, 271-284.

Mathai, A.M.(1967a). Dispersion and information. *Metron*, 26, 1-12.

Mathai, A.M.(1968). Some limit theorems in terms of dispersion. *Metron*, 27, 125-135.

Mathai, A.M. and Rathie, P.N.(1971). Characterizations of the concept of covariance and related measures in statistics through functional equations. *(Unpublished Report).*

Mathai, A.M. and Rathie, P.N.(1972). Characterization of Matusita's measure of affinity. *Ann. Inst. Statist. Math.*, **24**, 473-483.

Matusita, K.(1954). On the estimation by the minimum distance method. *Ann. Inst. Statist. Math.*, 5, 59-65.

Matusita, K.(1955). Decision rules based on the distance, for problem of fit, two samples and estimation. *Ann. Math. Statist.*, 26, 631-640.

Matusita, K.(1957). Decision rule based on the distance for the classification problem. *Ann. Inst. Statist. Math.*, 8, 67-77.

Matusita, K.(1961). Interval estimation based on the notion of affinity. *Bull. International Statist. Inst.*, 38, 241-244.

Matusita, K.(1967). On the notion of affinity of several distributions and some of its applications. *Ann. Inst. Statist. Math.*, 19, 181-192.

McGarthy, J. (1956). Measures of the value of information. *Proc. Nat. Acad. Sci. USA*, 42, 654-655.

McKean Jr. H.P.(1966). Speed of approach to equilibrium for Kac's caricature of a Maxwellian gas. *Arch. Rational Mech. Anal.*, 21, 343-367.

Muszély, Gy.(1973). On continuous solutions of a functional inequality. *Metrika*, 20, 65-69.

Natanson, I.P.(1964). *Theory of Functions of Real Variable. Vol.I.* Frederick Ungar Publishing Co., New York.

Nath, P.(1968). On the measures of errors in information. *J.Math.Sci.*, 3, 1-16.

Nath, P.(1970). Some axiomatic characterizations of a non-additive measure of divergence in information. *Symp. Information Measures. University of Waterloo, April 10-14, 1970.*

Ng, C.T.(1973). On the functional equation $f(x)+\Sigma_{i=1}^{n} g_i(y_i)=h(T(x,y_1,\ldots,y_n))$. *Ann. Polon. Math.*, 27, 329-336.

Ng, C.T.(1973a). On the measurable solutions of the functional equation $\Sigma_{i=1}^{2} \Sigma_{j=1}^{3} F_{ij}(p_i q_j) = \Sigma_{i=1}^{2} G_i(p_i) + \Sigma_{j=1}^{3} H_j(q_j)$. *(Unpublished Report).*

Ng, C.T.(1974). Representation of measures of information with a branching property. *Information and Control*, 25, 45-56.

Papaioannou, T. and Kempthorne, O.(1971). On statistical information theory and related measures of information. *Aerospace Research Laboratory Report*, No.ARL 71-0059.

Picard, C.(1965). *Théorie des Questionnaires*. Gauthier-Villars, Paris.

Pinsker, M.S.(1960). *Information and Information Stability of Random Variables and Processes (Russian)*. Izv. Akad. Mank. Moscow.

Pintacuda, N.(1966). Sul l'entropia di ordire α. *Annali Del l'Universita di Ferrara*, 12, 1-6.

Pintacuda, N.(1966a). Shannon entropy: A more general derivation. *Statistica*, 26, 509-524.

Quastler, H.(1953). *Information Theory and Biology*. University of Illinois Press, Urbana.

Quastler, H.(1956). *Information Theory in Psychology*. Free Press, Glencoe, Illinois (Proceedings of a conference held in 1954).

Rajagopalan, K.R.(1965). A note on entropy of Kannada prose. *Information and Control*, 8, 640-644.

Rao, C.R.(1949). On the distance between two populations. *Sankhya*, 9, 246-247.

Rao, C.R.(1954). On the use and interpretations of distance function in statistics. *Bull. Inter. Statist. Inst.*, 34, 90-97.

Rathie, P.N.(1970). On a generalized entropy and a coding theorem. *J. Appl. Prob.*, 7, 124-133.

Rathie, P.N.(1970a). On generalized measures of inaccuracies, information and errors in information. *Statistica*, 30, 340-349.

Rathie, P.N.(1971). On the solution of a functional inequality and its applications. *Tôhoku Math. J.*, 23, 681-690.

Rathie, P.N.(1971a). On generalized entropies, inaccuracies and informations. *Cahiers due Centre D'Etudes de Recherche Operationnelle*, 13, 98-105.

Rathie, P.N.(1971b). A generalization of the non-additive measures of uncertainty and information and their axiomatic characterizations. *Kybernetika*, 7, 125-132.

Rathie, P.N.(1971c). On some new measures of uncertainty, inaccuracy and information and their characterizations. *Kybernetika*, 7, 394-403.

Rathie, P.N.(1972). Generalized entropies in coding theory. *Metrika*, 18, 216-219.

Rathie, P.N.(1973). Some characterization theorems for generalized measures of uncertainty and information. *Metrika*, 20,122-130.

Rathie, P.N.(1974). Characterizations of the harmonic mean and the associated distance measure useful in statistical pattern recognition. *Symp. Measures of Information and Their Applications*, Indian Institute of Technology Bombay, August 16-18, 1974.

Rathie, P.N. and Kannappan, Pl.(1971). On a functional equation connected with Shannon's entropy. *Funkcialaj Ekvacioj*, 14, 153-159.

Rathie, P.N. and Kannappan, Pl.(1971a). On a characterization of inaccuracy. *Communicated for publication*.

Rathie, P.N. and Kannappan, Pl.(1972). A directed divergence function of type β. *Information and Control*, 20, 38-45.

Rathie, P.N. and Kannappan, Pl.(1973). An inaccuracy function of type β. *Ann. Inst. Statist. Math.*, 25, 205-214.

Rathie, P.N. and Kannappan, Pl.(1973a). On a new characterization of directed-divergence in information theory. *Trans. 6th Prague Conference on Information Theory, Statistical Decision Functions, Random Processes. September 19-25, 1971*, 733-745.

Rathie, P.N. and Nath, P.(1972). On inaccuracies, β-inaccuracies and errors in information. *Univ. Nac. Tucuman, Rev.Ser.A*, 22, 7-16.

Rényi, A.(1959). On a theorem of P.Erdős and its application in information theory. *Mathematica*, 1, 341-344.

Rényi, A.(1960). Az információelmelet néhány alapvető kérdése. *MTA III, Osztályának Közl.*, 10, 251-282.

Rényi, A.(1961). On measures of entropy and information. *Proc. Fourth Berkeley Symp. Math. Statist. and Probl.,1960*, University of California Press, 1961, Vol.1, 547-561.

Rényi, A.(1965). On the foundations of information theory. *Rev. Inst. Internat. Statist.*, 33, 1-14.

Schützenberger, M.P.(1954). Contribution aux applications statistiques de la théorie de l'information. *Publ. Inst. Statist. Univ. Paris*, 3, 3-117.

Shannon, C.E. (1948). A mathematical theory of communication. *Bell System Tech. J.*, **27**, 379-423, 623-656.

Sharma, B.D. and Ram Autar (1972). Generalized functional equation of two variables in information theory. *Revue Française d'Automatique, Informatique et Recherche Opérationnelle*, **6**, 85-98.

Sharma, B.D. and Ram Autar (1973). On characterization of a generalized inaccuracy measure in information theory. *J. Appl. Prob.*, **10**, 464-468.

Sibson, R. (1969). Information radius. *Z. Wahr. Verw. Geb.*, **14**, 149-160.

Siromoney, G. (1962). Entropy of logarithmic series distributions. *Sankhya Ser. A*, **24**, 419-420.

Siromoney, G. (1963). Entropy of Tamil prose. *Information and Control*, **6**, 297-300.

Siromoney, G. (1964). An informational-theoretical test for familiarity with a foreign language. *J. Psychological Researches*, **8**, 1-5.

Siromoney, G. and Rajagopalan, K.R. (1964). Style as information in Karnatic music. *The Journal of Music Theory*, **8**.

Theil, H. (1967). *Economics and Information Theory*. North-Holland, Amsterdam.

Toussaint, G.T. (1972). Some inequalities between distance measures for future evaluation. *IEEE Trans. Computers*, C-21, 409-410.

Toussaint, G.T. (1973). Distance measures as measures of certainty and their application to statistical pattern recognition. (Presented at the First Conference of the Statistical Science Association of Canada held at Queen's University, Kongston, June 4-6, 1973).

Toussaint, G.T. (1974). Sharper lower bounds for discrimination information in terms of variation. *IEEE Trans. Information Theory* (To Appear).

Toussaint, G.T. (1974a). On the divergence between two distributions and the probability of misclassification of several decision rules. *Proc. Second International Joint Conference on Pattern Recognition, August 13-15, 1974, Copenhagen.*

References

Toussaint, G.T. (1974b). Some properties of Matusita's measure of affinity of several distributions. *Ann. Inst. Statist. Math., (To Appear)*.

Toussaint, G.T. (1974c). Discrimination and the affinity of several distributions. *Presented at the Conference of the Statistical Science Association of Canada, May 30- June 1,1974, Toronto*.

Toussaint, G.T. (1974d). On some measures of information and their applications to pattern recognition. *Proc. Symp. Measures of Information and Their Applications*. Indian Institute of Technology Bombay, August 16-18, 1974.

Tverberg, H. (1958). A new derivation of the information function. *Math. Scand.*, **6**, 297-298.

Vajda, I. (1968). Axiomy a-entropie zobecněného pravděpodobnostního schématu. *Kybernetika*, **4**, 105-112.

Vajda, I. (1970). Note on discrimination information and variation. *IEEE Trans. Information Theory*, IT-16, 771-773.

Varma, R.S. (1966). Generalizations of Rényi's entropy of order α. *J. Math. Sci.*, **1**, 34-48.

Volkonskij, V.A. and Rozanov, Ju.A. (1959). Some limit theorems for random functions I. *Theory of Probability and Applications (USSR). English Translation*, **4**, 178-197.

LIST OF SYMBOLS

$\Sigma = \Sigma_{i=1}^{n}(.)$, 1

$S_n = \{P=(p_1,\ldots,p_n), p_i \geq 0, \Sigma p_i=1\}$, 2

$S_n^* = \{P=(p_1,\ldots,p_n), p_i > 0, \Sigma p_i=1\}$, 2

ε - element of, 2

[.,.] - closed interval, 2

$H_n(p_1,\ldots,p_n) = -\Sigma_{i=1}^{n} p_i \log p_i$, 2

$\{a_1,\ldots,a_n\}$ - set of numbers $a_1,\ldots,a_n$, 3

$|a|$ - absolute value of a, 5

[.,.[- right open left closed interval, 7

].,.] - left open right closed interval, 7

].,.[- open interval, 7

D^+f, D^-f - the right and left derivatives, 11

$\mathcal{R}$ - the real line, 11

$\hat{H}_{n,\alpha}$ - additive entropy of order α, 12

$H_{n,\alpha}$ - non-additive entropy of order α, 12

$\cup$ - union, $\cap$ - intersection of sets, 19

$\to$ - tends to, 19

$\mu(.)$ - Lebesgue measure, 19

$H_{n,\alpha}^{\beta_i}$, $\hat{H}_{n,\alpha}^{\beta_i}$ - generalized entropies, 22

$H_n(p_1,\ldots,p_n;\beta)$, $I(u_1,\ldots,u_n;p_1,\ldots,p_n)$ - generalized entropies, 23

inf - infimum, 31

$I_n(p_1,\ldots,p_n;q_1,\ldots,q_n) = I_n(P:Q)$ - directed divergence between two discrete distributions, 35

$\hat{I}_{n,\alpha}(P:Q)$ - additive directed divergence of order α, 46

$I_{n,\alpha}(P:Q)$ - non-additive directed divergence of order α, 46

$I_{n,\alpha}$, $I_{n,\alpha}^{\beta}$, I_n^{β} - generalized directed divergences, 54

$I_n(P:Q:R)$ - directed divergence involving three distributions, 55

$I_{n,\alpha}(P:Q:R)$, $\hat{I}_{n,\alpha}(P:Q:R)$ - generalized directed divergences of order α, 56

List of Symbols

$J_n(p_1,\ldots,p_n:q_1,\ldots,q_n)$ - *Pseudo-measure of directed divergence*, 61

$(\ )\times(\ )$ - *product space*, 70

$H_n(P:Q)$ - *inaccuracy*, 75

$H_{n,\alpha}(P:Q)$ - *non-additive inaccuracy of order* α, 80

$\hat{H}_{n,\alpha}(P:Q)$ - *additive inaccuracy of order* α, 80

$H_\alpha(p_1,\ldots,p_n:\beta_1,\ldots,\beta_n:q_1,\ldots,q_n)$ - *generalized inaccuracy*, 84

$Cov(x,y)$ - *covariance between* x *and* y, 88

$E(.)$ - *mathematical expectation of* $(.)$, 88

$Var(x)$ - *variance of* x, 90

$\cos\theta$ - *cosine of* θ, 91

$\rho_n(p_1,\ldots,p_n:q_1,\ldots,q_n)$ - *affinity between* $(p_1,\ldots,p_n)$ *and* $(q_1,\ldots,q_n)$, 91

$D(X)$ - *dispersion in* X, 97

$||C||$ - *norm of the matrix* C, 99

$P*Q$ - *the set of elements* $p_i q_j$, $i=1,\ldots,n$, $j=1,\ldots,m$, 115

AUTHOR INDEX

Aczél,J., 1,10,12,22,31,39,45,59,
    88,105,107,108,109,115,
    119.
Adke,S.R., 100,119.
Ali,S.M., 101,119.
Arlotti,L., 105,117,120.
Ash,R., 24,25,69,120.
Attneave,F., 28,120.
Autar,R., 129,130.
Baiocchi,C., 120.
Baker,J.A., 59,119.
Balasubrahmanyam,P., 27,120.
Beckenbach,E.F., 120.
Belis,M., 23,120.
Bellman,R., 120.
Bending,A.W., 24,120.
Bhargava,T.N., 27,102,120.
Bhattacharyya,A., 91,111,120.
Blachman,N.M., 120.
Borges,R., 1,30,120.
Brillouin,L., 24,120.
Campbell,L.L., 1,25,26,31,35,67,
    70,73,120,121.
Charvát,F., 1,12,32,123.
Chaundy,T.W., 1,8,121.
Cover,T.M., 114,121.
Cozzolino,J.M., 27,121.
Csiszár,I., 24,67,111,121.
Daróczy,Z., 1,14,22,28,30,31,32,
    78,105,107,108,115,
    119,121,122.
Doyle,P.H., 27.
Djoković,D.Ž., 59,119.
Dutta,M., 24,67,122.
Erdös,P., 65,122.
Faddeev,D.K., 1,6,29,122.
Fano,R.M., 68,122.
Feinstein,A., 25,122.

Fischer,P., 45,122.
Forte,B., 1,10,12,15,30,32,119,
    122.
Gallager,R.G., 24,25,123.
George,A., 102,123.
Gini,C., 23,101,123.
Good,I.J., 123.
Gottinger,H.W., 114,123.
Guiasu,S., 23,120.
Hart,P.E., 114,121,123.
Hartley,R.V.L., 2,23,123.
Havrda,J., 1,12,32,123.
Hellman,M.E., 114,123.
Hewit,E., 19,123.
Hobson,A., 67,123.
Hoeffding,W., 112,123.
Horibe,Y., 114,123.
Horowitz,I., 27,123.
Hudimoto,H., 114,124.
Hyvärinen,L.P., 27,124.
Ikeda,S., 112,124.
Ito,T., 114,124.
Jaynes,E.T., 67,124.
Jeffreys,H., 100,124.
Jelinek,F., 1,26,27,124.
Kailath,T., 113,114,124.
Kannappan,Pl., 1,18,20,21,31,35,
    37,48,53,58,59,61,67,
    70,71,72,73,75,82,84,
    85,86,110,119,124,125,
    129.
Kapur,J.N., 22,54,125.
Katai,I., 1,28,31,122,125.
Kaufman,H., 75,82,87,96,102,105,
    107,108,116,125,126.
Kempermann,J.H.B., 111,126.
Kempthorne,O., 128.
Kendall,D.G., 1,28,29,126.

Author Index

Kerridge,D.F., 68,75,83,84,85,126.
Khinchin,A.J., 1,6,126.
Kirmani,S.N.U.A., 90,126.
Kullback,S., 24,67,68,111,126.
Lee,P.N., 1,28,29,126.
Leibler,R.A., 126.
Mallows,C.L., 84,126.
Mathai,A.M., 87,96,97,102,123,125, 126,127.
Matusita,K., 90,102,113,114,127.
McGarthy,J., 127.
McKean,H.P., 111,127.
McLeod,J.B., 1,8,121.
Muszély,Gy., 45,127.
Natanson,I.P., 127.
Nath,P., 84,105,109,115,119,127, 129.
Ng,C.T., 1,10,12,15,16,18,21,35, 71,72,82,119,122,124, 127.
Papaioannou,T., 128.
Pfanzagl,J., 45,119.
Picard,C., 25,128.
Pinsker,M.S., 111,128.
Pintacuda,N., 1,29,120,122,128.
Quastler,H., 24,68,85,128.
Rado,F., 59,119.
Rajagopalan,K.R., 27,128,130.
Rao,C.R., 102,128.
Rathie,P.N., 1,20,22,23,26,31,32, 35,37,42,45,48,53,54, 58,59,61,67,70,71,72, 73,75,82,84,85,87,102, 105,107,108,109,110,116, 124,125,126,127,128,129.
Raviv,J., 114,123.
Rényi,A., 1,12,29,105,106,107,115, 116,129.
Rozanov,J.A., 111,131.
Schützenberger,M.P., 129

Shannon,C.E., 1,2,4,6,130.
Sharma,B.D., 130.
Sibson,R., 68,130.
Siromoney,G., 27,120,130.
Silvey,S.D., 101,119.
Stromberg,K., 19,123.
Theil,H., 27,56,68,69,130.
Toussaint,G.T., 109,112,113,114, 130,131.
Tverberg,H., 1,6,28,131.
Uppuluri,V.R.R., 27,102,120.
Vajda,I., 105,107,111,115,131.
Varma,R.S., 22,131.
Volkonskij,V.A., 111,131.
Wolfowitz,J., 112,123.
Zahner,M.J., 27,121.
Zappa,G., 123.

SUBJECT INDEX

Affinity, 90
- among several distributions, 95
- Bhattacharyya's coefficient of, 91,111
- characterization, 91

Association measure, 101
Binary alphabet, 25
Bit, 2
Classification, 99
Coding theory, 25
Coefficient of divergence, 101
Communication theory, 25
Concentration ratio, 101
correlation,
- linear, 90
- multiple, partial, serial, inter-class, intra-class, 101
- ratio, 101

Cost function, 25
Covariance, 88
- characterization, 88

Decision function, 100
Directed divergence, 35
- additive, 46
- applications, 67
- characterizations, 37,48,61
- continuous analogues, 67
- function, 36,46
- functional equation, 42
- generalized, 54,55,56
- non-additive, 46
- of order α, 46
- pseudo-measures, 61

Discrepancy, 96
Discriminating power, 85
Discrimination, 99

Dispersion, 97
- analysis, 97
- joint, multivariate, 98
- generalized, 98,99
- principle of minimum, 100

Distance
- angular, 102
- between populations, 99
- continuous analogues, 100
- generalized, 104
- Kolmogorov's variational, 100,111
- Matusita's, 95

Distributions
- cumulative, 104
- generalized, 106
- incomplete, 105
- multinomial, 99
- union of, 105

Diversity index, 23, 101
Ecology, 27
Eigen value, 99
Entropy, 2
- additive, 12
- applications, 24
- characterizations, 4,14
- continuous analogues, 23
- function, 13
- functional equation, 17
- generalized, 22
- non-additive, 12
- of order α, 12

Estimation
- minimum dispersion, 100
- minimum risk, 101
- point and interval, 102

Estimator, 100

Subject Index

Equivocation of inference, 84
Expectation, mathematical, 88
f-divergence, 67
Gini's mean difference, 101
Goodness-of-fit, 96
Harmonic mean, 109
Holder's inequality, 26
Inaccuracy, 75
- additive, 80
- applications, 84
- characterizations, 77,82
- continuous analogues, 84
- function, 76,80
- generalizations, 84
- non-additive, 80
- of order α, 80
- pseudo-measures, 83

Industrial concentration, 27
Information, 1
- apparent, 68
- content, 25,27,68
- discrimination, 68,110
- Fisher's, 101
- gain, 67,68
- improvement, 69
- inaccuracy, 68
- measure, 23
- radius, 68

Jeffreys' invariant, 100
J-divergence, 111
Kurtosis, 101
Likelihood function, 101
Mean absolute deviation, 98
Metric, 97
Metrix, 98
Moments
- absolute, 98
- central, 101

Nit, 2
Noiseless coding theorem, 25
Norm, 99
Parametric space, 104
Pattern recognition, 109
Questionnaire theory, 24
Random variable, incomplete, 106
Range, interquartile, 101
Risk, 100
Skewness, 101
Stimuli, 27
Utility, 23
Variance, 90

Q
360
M322

APR 27 1978